Revelations of Chernobyl
by Nakasuji Jun

廃墟
チェルノブイリ

写真・文 中筋 純

二見書房

リアル～まえがきにかえて

　2007年晩秋。実りの季節。小振りな実をたわわに付けたリンゴの木が風に揺られている。思わず手を伸ばしそうになるが、ウクライナ人曰くこのリンゴは「禁断のリンゴ」、間違って食べようものなら楽園を追われるのは必至なのだそうだ。

　1986年4月26日、旧ソ連ウクライナ共和国のチェルノブイリ原子力発電所4号炉が爆発炎上、炉心のメルトダウンを引き起こし大量の放射性物質が大気中に放出された。22年経った現在でも、発電所周辺はもとより隣国ベラルーシなどで放射性物質による水質土壌汚染は解決されておらず、幾種類かの放射性物質は地中深くに根ざし植物に吸収されている。この地に生えるリンゴは22年前から「毒リンゴ」という不名誉なレッテルを貼られてしまった。

　しかし人間が消滅せぬ放射能の恐怖に及び腰なのに対し、自然の再生は急ピッチだ。天敵人間がいなくなったことで、ウクライナ平原本来の自然のリズムや掟が取り戻されているかのようである。その混沌とした状況の中で、発電所の巨大な煙突はいまだに不気味な輝きを発し、背後に控えるゴーストタウン、プリピアチとともに負の存在感をあらわにしている。この光景はいまだに地球を食い尽くそうとしている人間の未来に対する「黙示録」かもしれない。

　バーチャルな世界が人間の頭を支配し始めている昨今、廃墟という存在は訪れる人間に紛れもないリアルを突きつける刃のような存在だと僕は思っている。未だ鉄条で囲まれ人間のみが立ち入ることを許されない、世界でも希有な場所チェルノブイリ。大自然に飲み込まれつつあるその亡骸が語るものも紛れもないリアル、行き先がわからず混沌に身を任せざるをえない我々人類の現状に対するリアルだ。

Preface-"Real"

It was late autumn - harvesting season. Apple trees, which were heavy with small apples, were swaying in the wind. When I was about to reach my hand to an apple to take it, a Ukrainian man said, "Don't touch it. This apple is not edible. If you eat it by mistake, you will be purged from the earthly paradise".

On April 26, 1986, at the Chernobyl nuclear power plant No.4 reactor following the meld down of a reactor core, graphite caught fire, and a large quantity of radiation was observed in the air. Even now, though 22 years have passed since then, fearful radiation absorbed in the soil is still detected in the vicinity of the No.4 reactor but also in Ukraine Republic of defunct Soviet Union. Many kinds of radioactivity soaked into the ground are absorbed in the vicinity of the plants. For this reason, the apples produced in this district are given disgraceful name of 'Adam's Apple'.

But compare with human-beings that fear eternal existence of radioactivity, nature's regeneration is going on actively. It is as if iron laws of nature at the Ukrainian plain itself recovered without enemy called Nature ≡ human-beings. In the chaotic situation, the huge, tall chimneys still towering into the sky and the ghost town of Pripiati over there showed the existence of negative. This view may be a relation of humankind which might try to eat up the asset of the world in the future.

In these days when virtual world is trying to control human's brain, I believe that the existence of ruins is like testimony which shows us the existence of obvious facts. The place named Chernobyl is a rare place in the world where only mankind is not allowed to enter being surrounded with barbed wire fence. Chernobyl is under the control of nature without mankind, and its ruins tell us the reality of the world itself ≡ the reality of our human-beings which only have to resign itself to the chaos knowing nowhere to go.

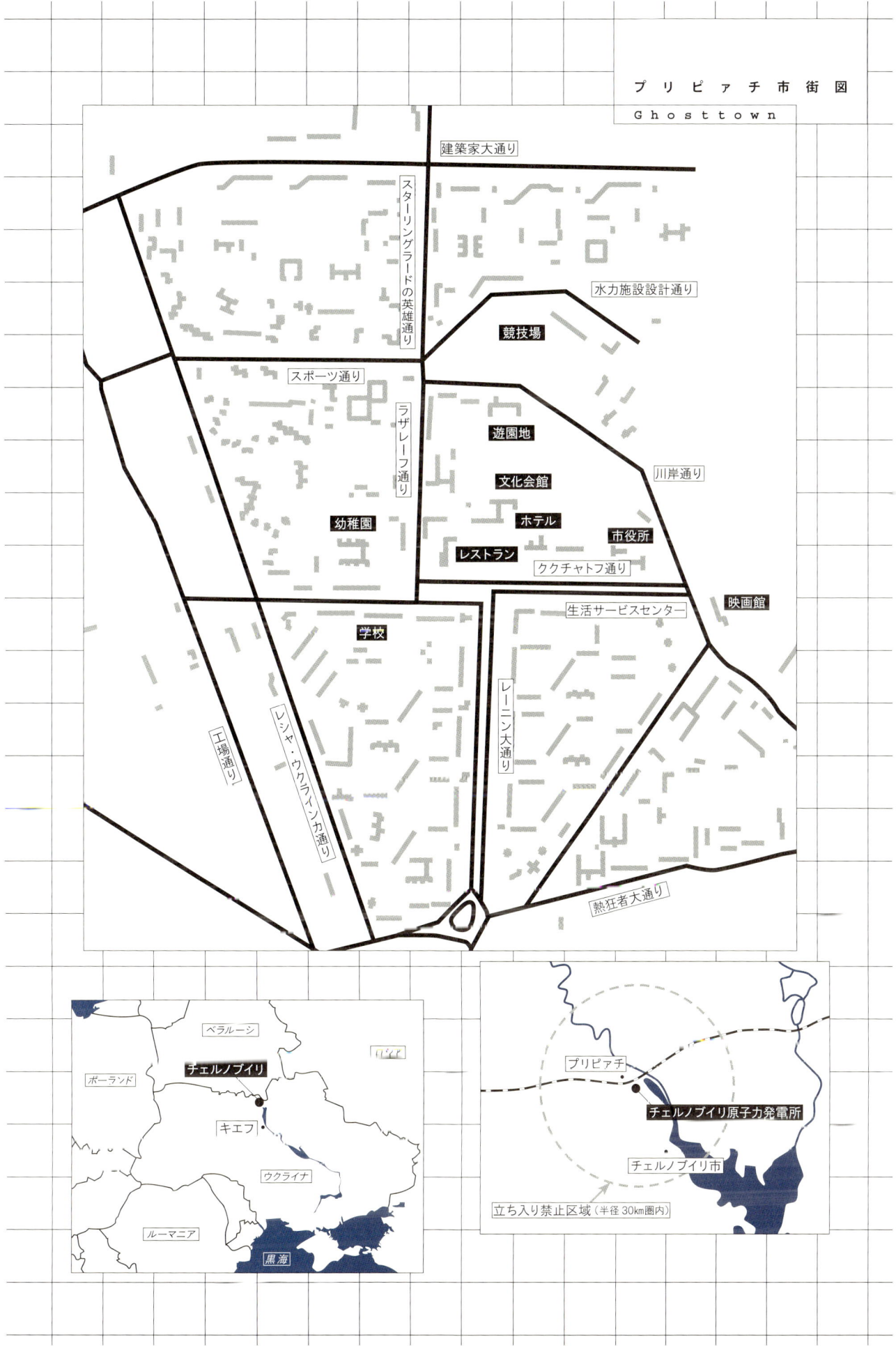

失われたユートピア～プリピアチ　1

　発電所の北西約3キロにチェルノブイリ原子力発電所の城下町的存在プリピアチがあった。ソ連邦の未来を支える原子力事業の根幹である発電所の職員とその家族約5万人が、当時のソ連にしては高水準の暮らしを日々送っていた。衣食住に関連するすべての施設はおろか、学校、病院などの生活施設、音楽ホールや遊園地などの余暇施設までも兼ね備えた密集型高機能都市は、ソ連人民のあこがれの町、ソ連国家の理想を具現したユートピア都市であった。

　しかしユートピアは一瞬にしてゴーストタウンへと姿を変える。爆発を起こし瓦礫と化した4号炉から放たれた広島型原爆の500倍ともいわれる放射能が、瞬時に町を覆ってしまったのはいうまでもない。事実を隠しきれなくなった当局は、事故後36時間経ってようやく全住民に一時避難命令を出した。だが3日分の手荷物を持ってバスに詰め込まれた住民たちが二度とこの町に戻ってくることはなかった。

　それから22年、人間のユートピアにはただただ崩落の時期を待つ武骨なコンクリートアパートメントの廃墟が林立するだけの空間となった。皮肉なことに人間が排除されたあとは、野生動物が我が物顔で闊歩し剪定されないポプラの街路樹は町を覆い尽くさんばかりに成長している。黄色く染まった葉が晩秋の冷たい風に吹かれてかさこそ鳴く様は、まさに悲運な町へのゴスペルのようだ。息を切らしながら町最大の16階建てマンション廃墟の屋上を目指す。目に飛び込んでくるものは、果てしなく続くウクライナの大平原と、役目を終え静かに木々の大海に飲み込まれるユートピア、そして未だ不気味な威容を放つ発電所の石の棺桶だ。その光景はまさに人類が築き上げた近代文明の墓碑銘である。

The lost Utopia, Pripiati 1

One of the dormitory towns, Pripiati was about 3 km northwest from Chernobyl nuclear power plant. The staffs of power plant and their family, about 50,000 people were living there provided high level facilities, not only for just making a living but also schools, hospital, music hall, amusement park, and so on. This high qualified city was a Utopian city which embodied an ideal of the Soviet Union.
But, this town collapsed to a ghost town in a moment. The large amount of radioactivity which released from No.4 reactor was about 500 times more of Hiroshima's. The spread of radiation released by the accident had been observed all over the city. At first the Soviet Union considered this accident as a mere domestic problem and evaded the official announcement. It was after 36 hours that the announcement concerning the accident was released officially.
22 years after the accident, the Utopian city has been transformed to a ruined town where only a great number of collapsed apartments exist. After people left the town, ironically wild animals are walking around without fear of radioactive and the roadside poplar trees are growing as if they were concealing the ruin city. The sight yellow turned leaves rattling in the breeze sounded like a gospel for the town. I stepped up to the top of the 16-storied apartment house panting. What I found there were the boundless Ukrainian plain, the Utopia was swallowed in the sea of trees, and the power plants were like a huge stone coffin. That sight was the very epitaph of modernism which humankind built.

The Ferris wheel in Paradise Lost

When spring comes on Ukrainian plain in late April, the warm temperature melts the frozen field down and it begins to breeze from the south. Birds sing, cherry tree blossoms bloom, and children play outside taking off thick sweater. Workers at the power plant are looking forward to May Day as one of the spring festival.

Above all, here in Pripiati, there was big news that year. It was an opening of the amusement park near central park. Especially it was a Ferris wheel. From there people would be able to command the whole town. To say nothing of children and couples, even elderly persons leaped their hearts, I was told. This park was to be a small paradise in the communist country where only regulation controlled the society. However, nobody was in the town on May Day that year. The town was exposed to radioactivity because of the power plant accident that had occurred only 5 days before May Day. The Ferris wheel has been exposed to wind and rain for 22 years. The end of October when I visited this place was the season Ukrainian plain was dyed to madder red. Orange colored rusty Ferris wheel was melted into the trees completely turned red under the clear sky. Chilly wind in late fall occasionally shook vacant gondolas, making creaky sound.

It did not become a paradise, but turned to be a Paradise Lost of today. The Ferris wheel of Pripiati which froze peoples' expectations is going on swinging as the monument of nuclear pollution today.

失楽の観覧車

　ウクライナ平原の4月の後半といえば長い冬から解放されすべてのものが春の空気に包まれ始める、1年でも最も躍動にあふれる季節だそうだ。小鳥がさえずり、桜がほころび、たくさんの子供たちが厚手のセーターを脱ぎ捨てて外で遊び回る。そして発電所で働く人の心待ちは、なんといっても5月1日のメーデーである。労働者のお祭りはここでは春祭りも兼ねていた。

　プリピアチのその年のメーデーにはなによりもビッグなニュースがあった。中央広場の裏手に建設が進んでいた遊園地が開園するからだ。空中ブランコやゴーカートなど遊具は本当にささやかだが、なかでも目玉は小さいながらも町を見渡せる観覧車であった。この観覧車に子供たちはもちろん、若いカップルや、老人までもが胸躍らせていたそうである。娯楽が少ない殺伐とした共産国家小さな楽園の誕生のはずだった。

　だが、メーデーを迎える頃には町には誰もいなくなった。小さな楽園は開園のわずか5日前に起きた発電所事故で放射能汚染され、人々の嬌声に1度も包まれることをしらずに22年間もの間、風雨にさらされ続けている。僕がこの場所を訪れた10月下旬はウクライナ平原が茜色に染まる季節だ。赤錆が浮いたオレンジ色の観覧車は、澄み渡った青空の下燃え盛る木々にとけ込んでいる。時折吹く晩秋の冷気を含んだ風は無人のゴンドラをわずかに揺らし、金属のきしんだ音が辺りに響き渡る。いまにも動き出しそうだ。

　楽園にならずして失楽園へ。プリピアチの観覧車は人々の期待をフリーズさせたまま、核汚染のモニュメントとして廃墟の町で今日も風に揺られている。

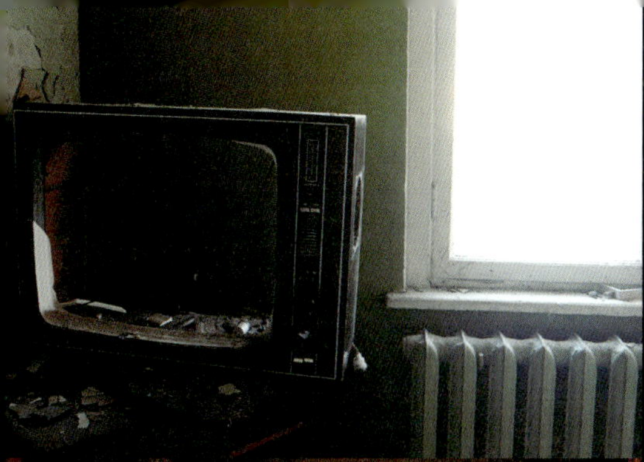
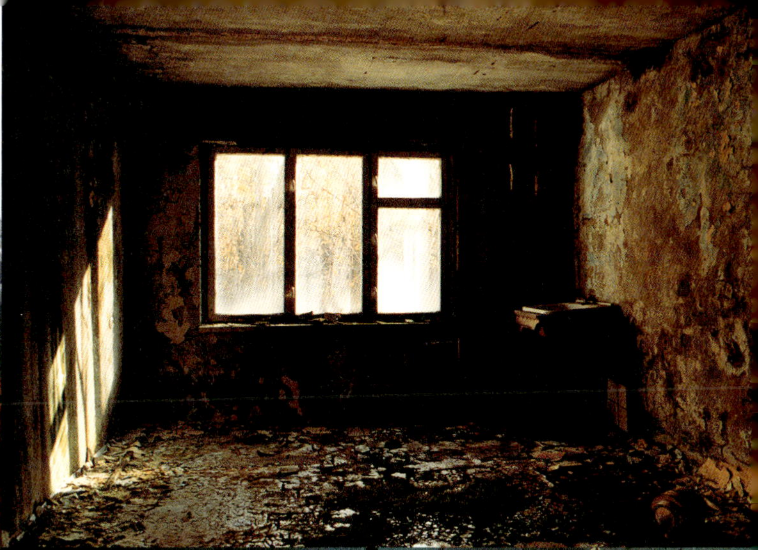
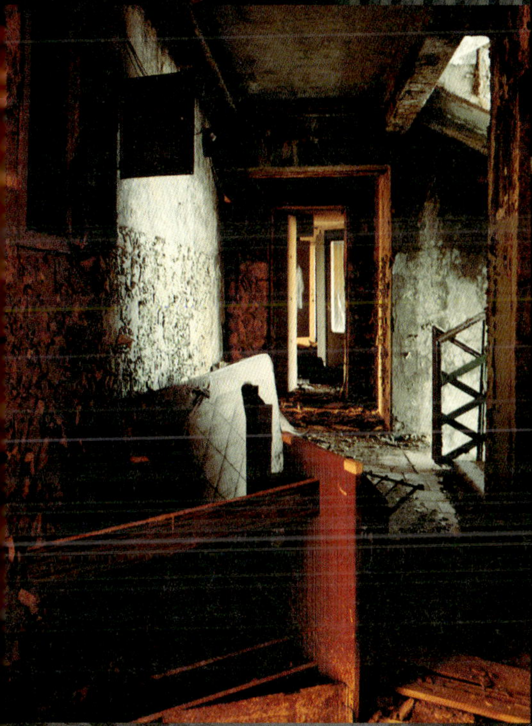
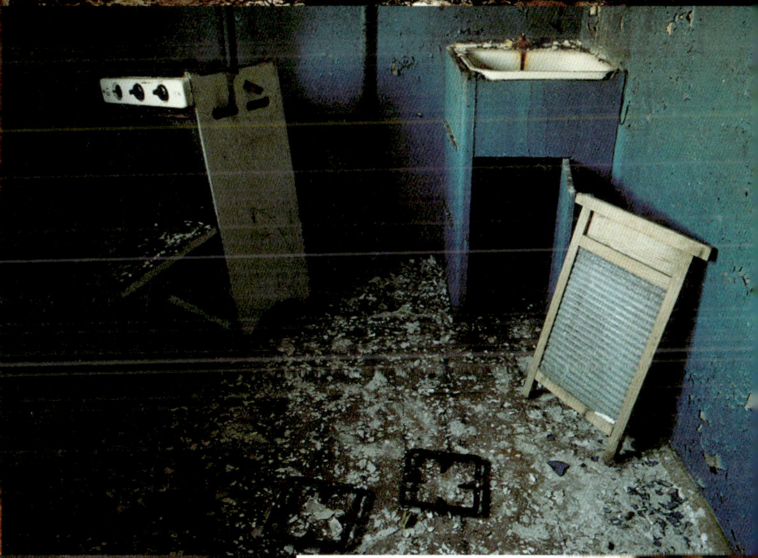
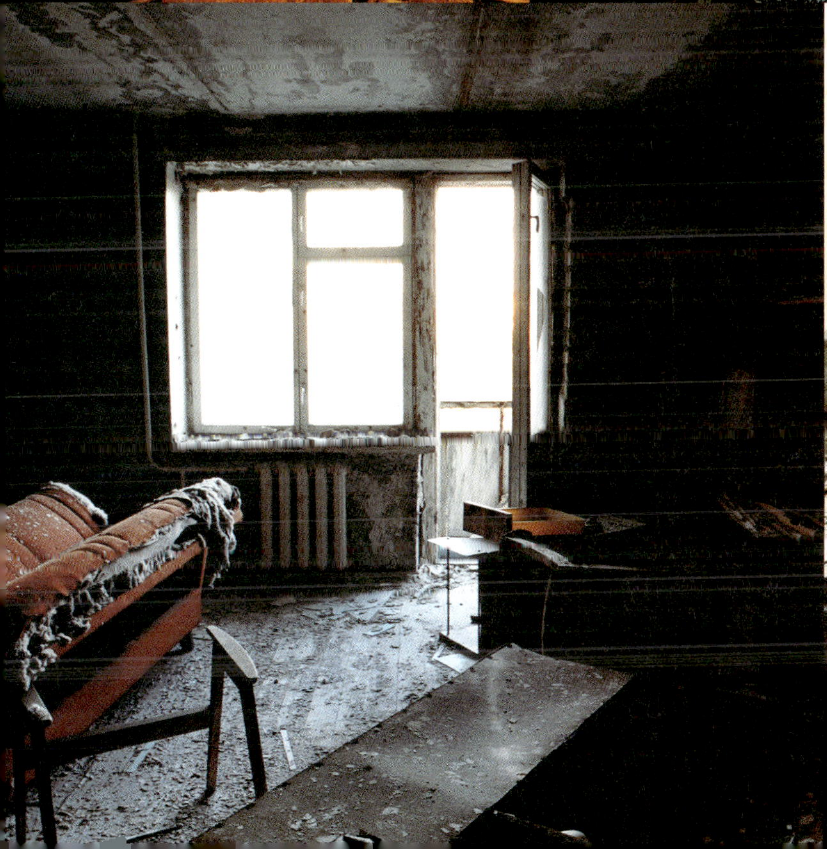
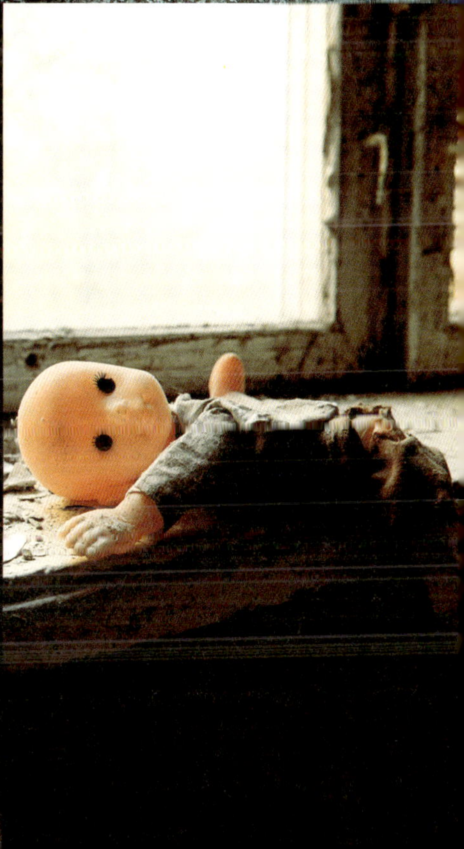

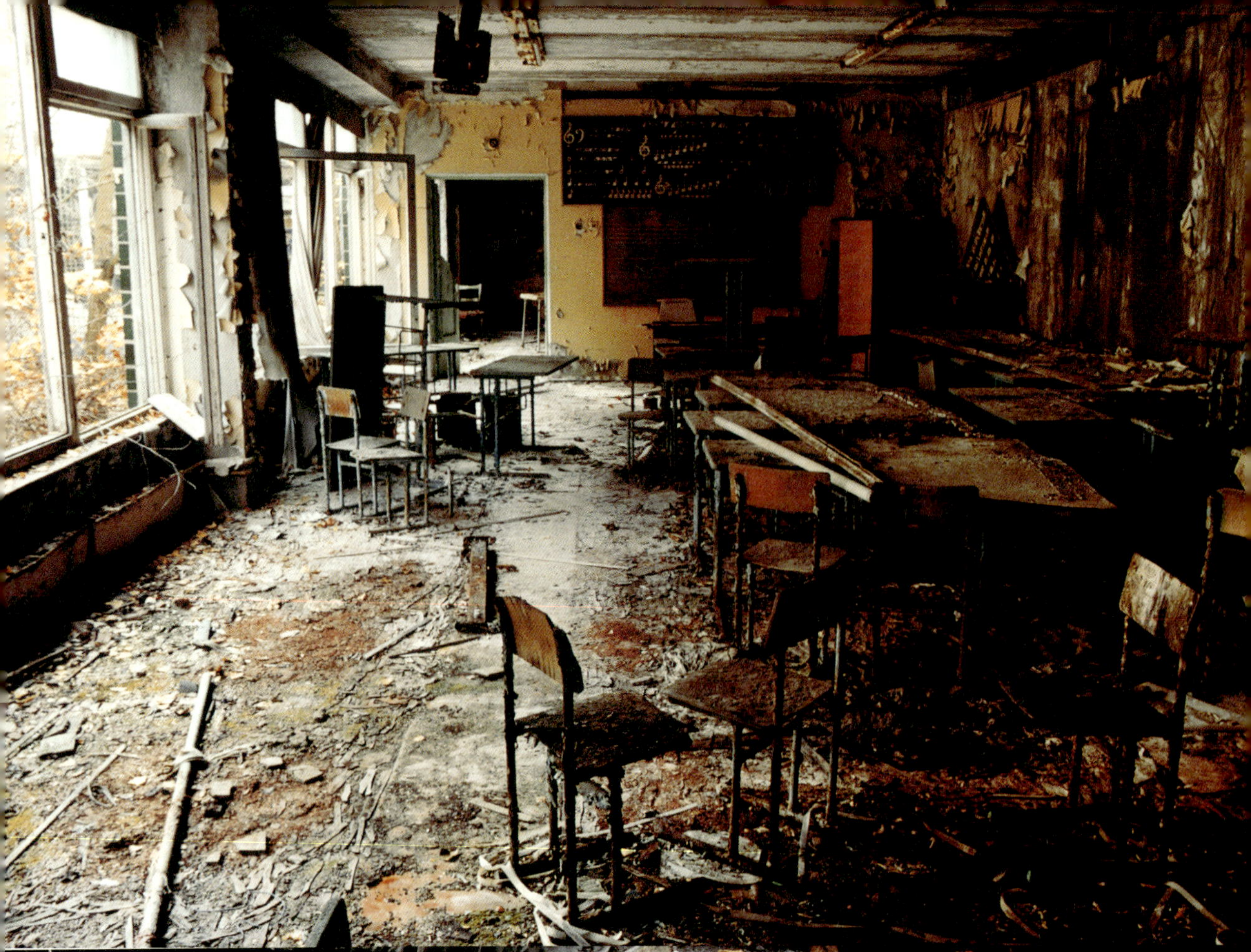

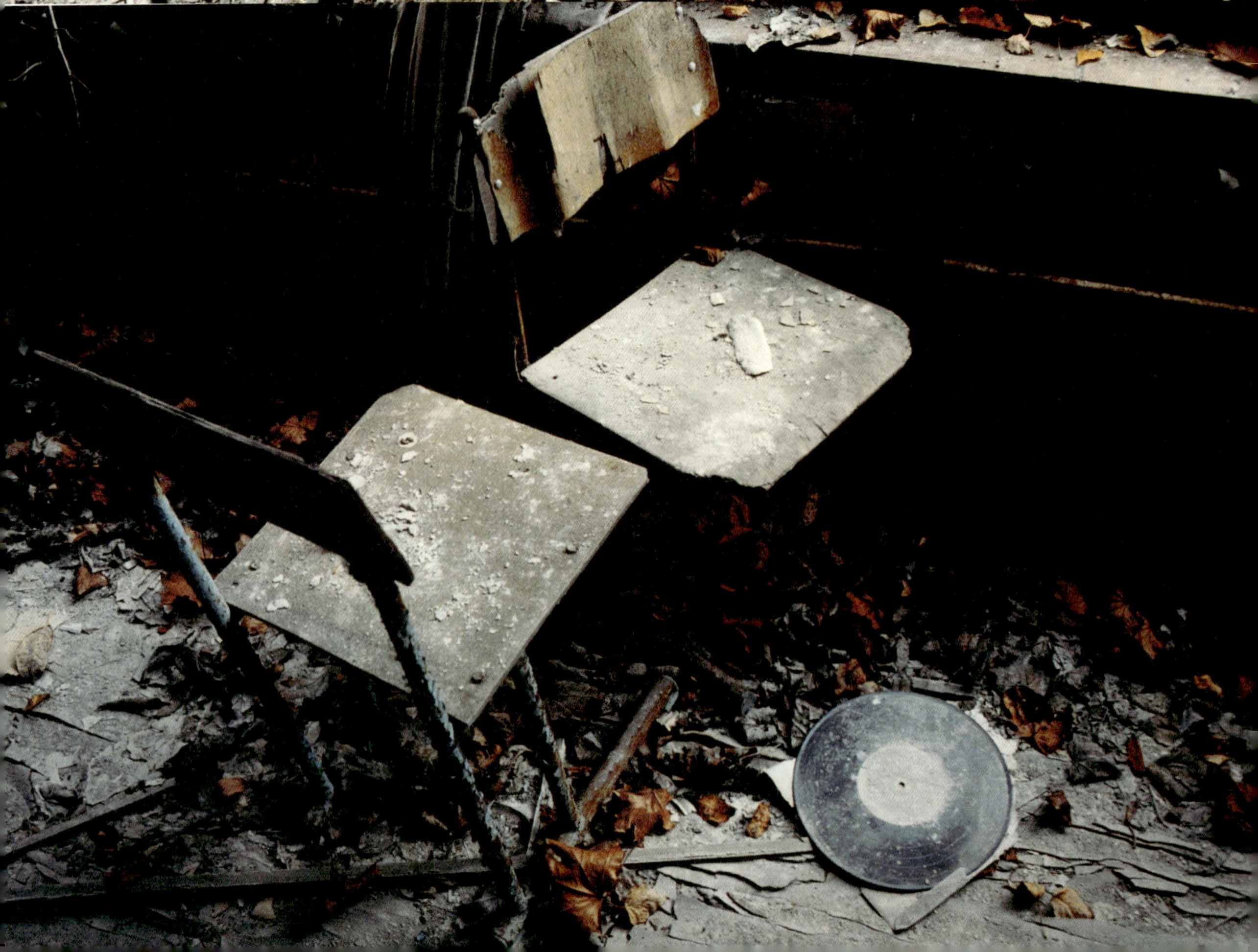

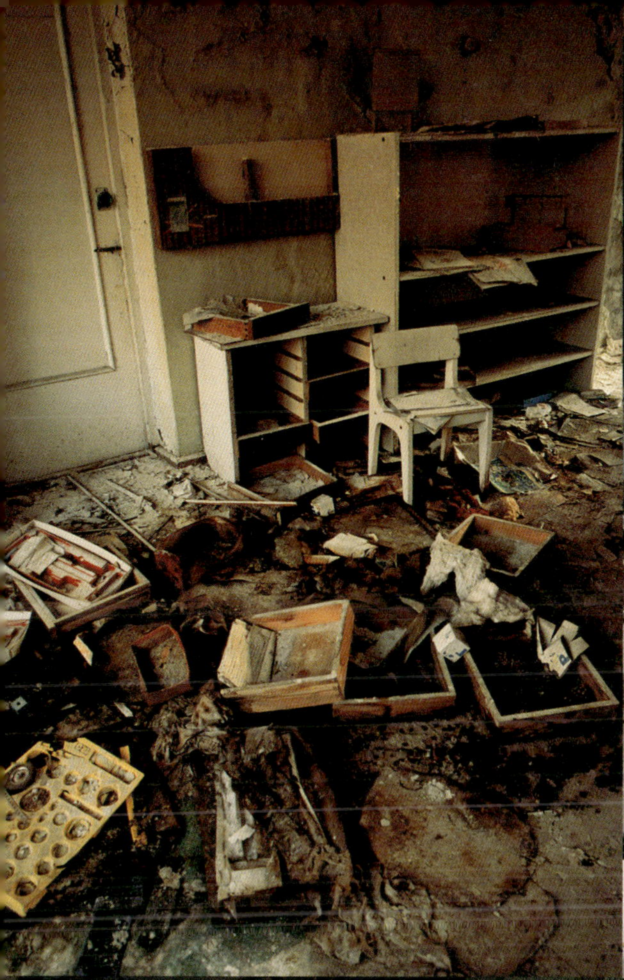

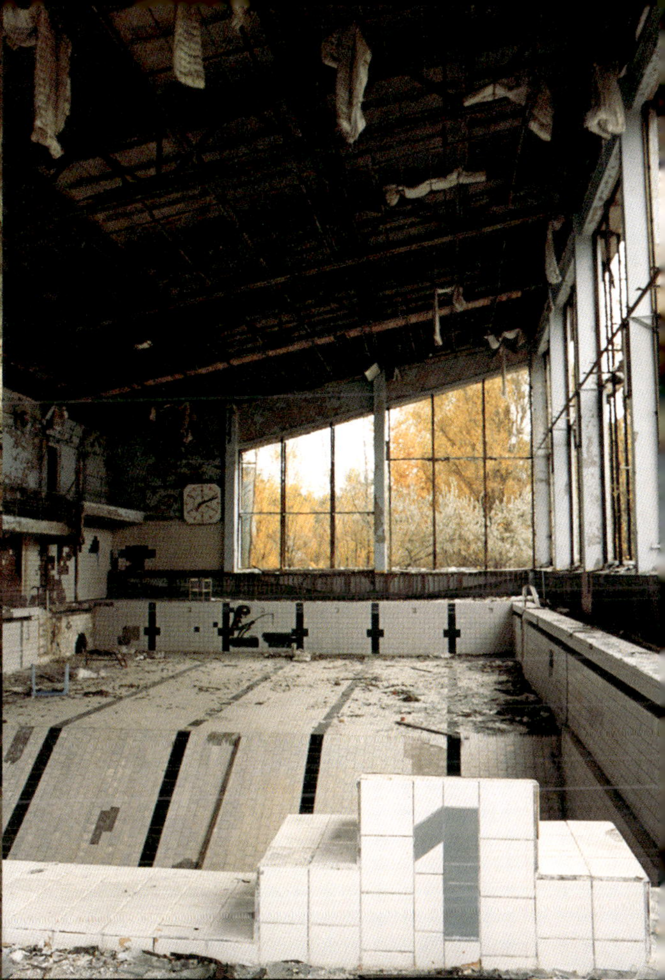
1

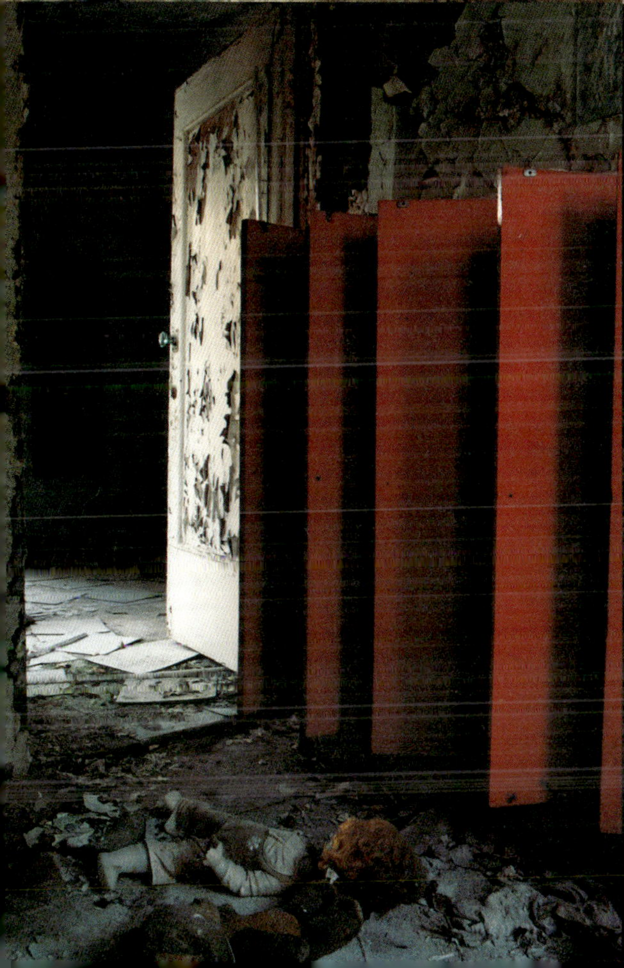

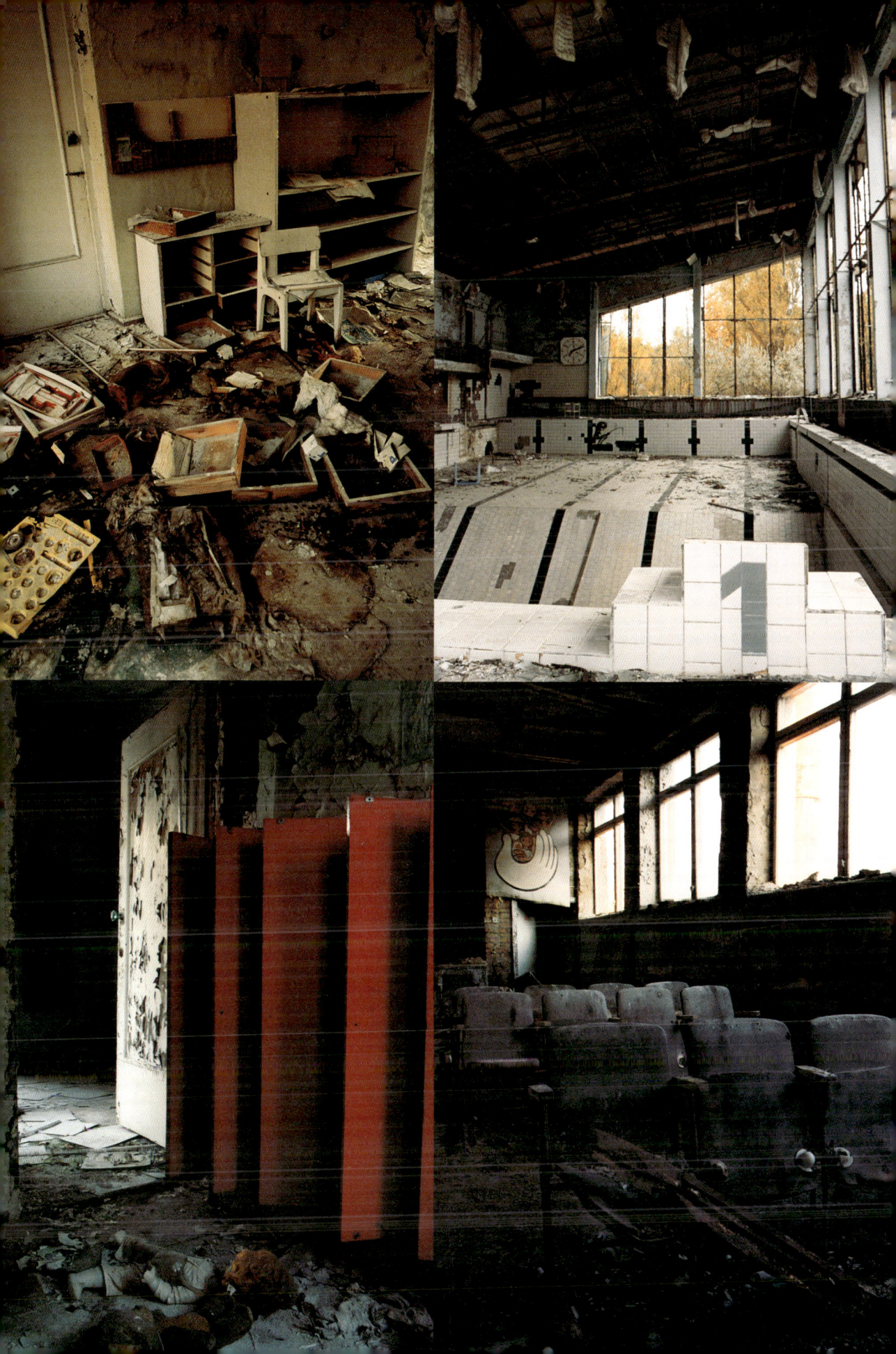

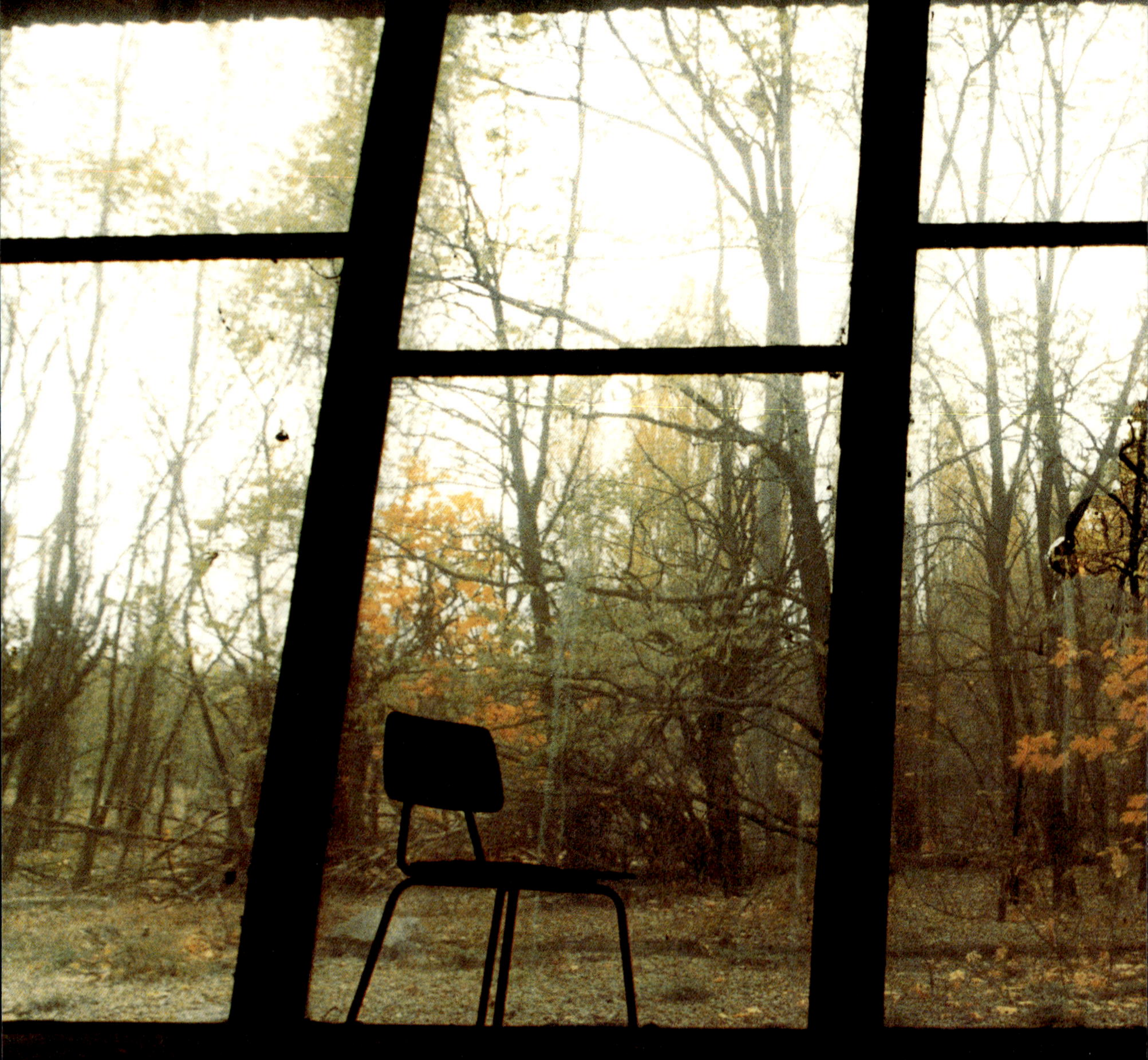

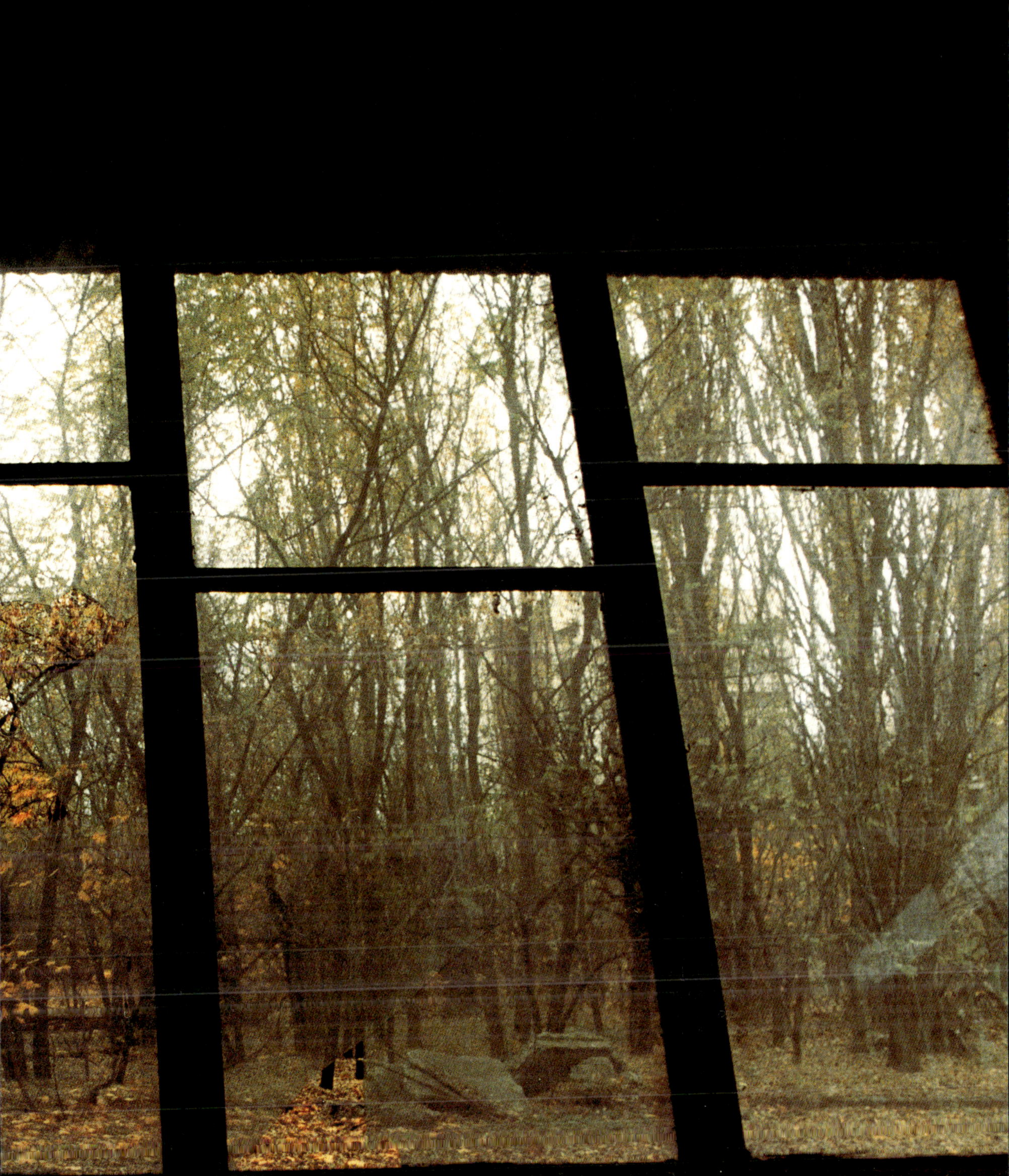

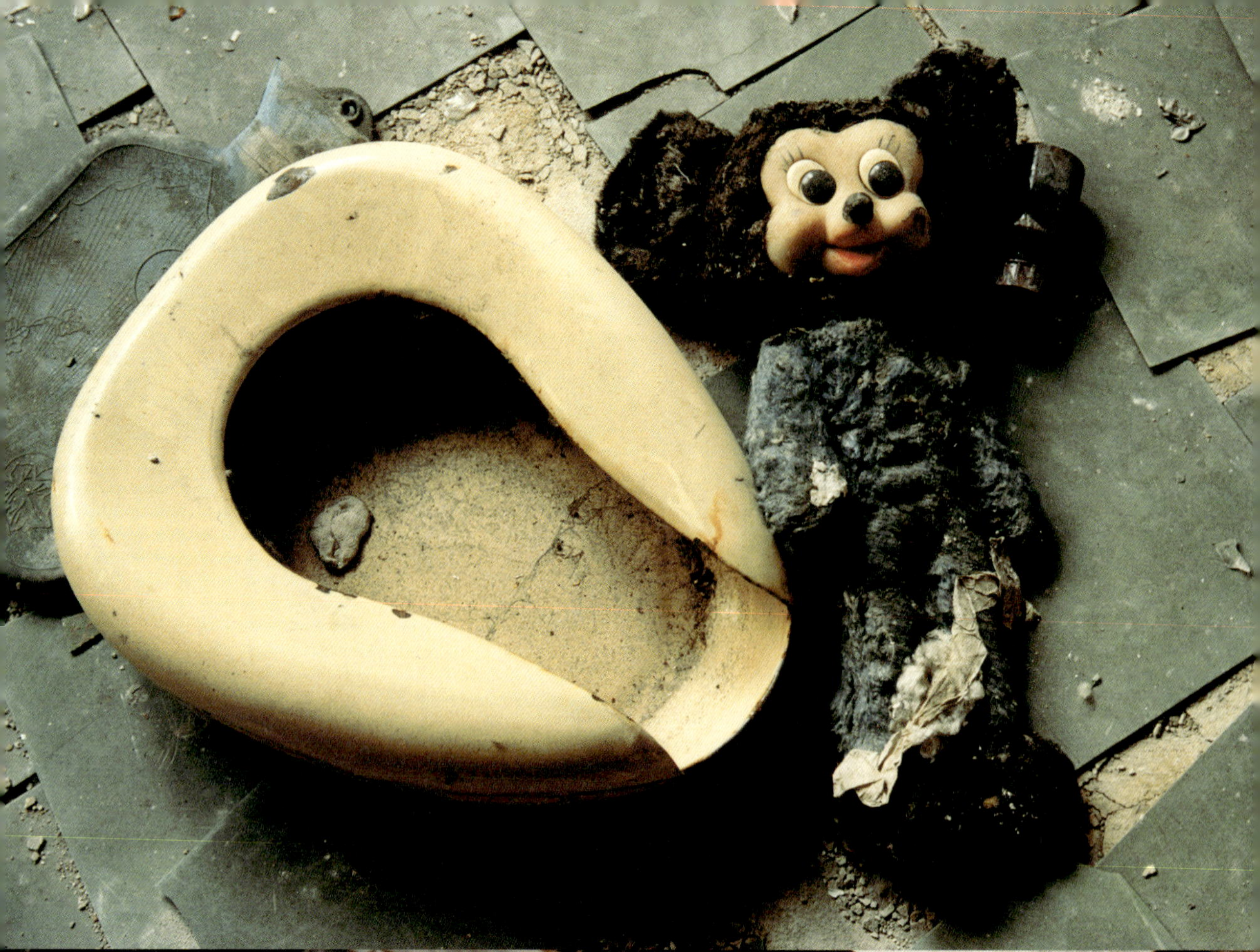

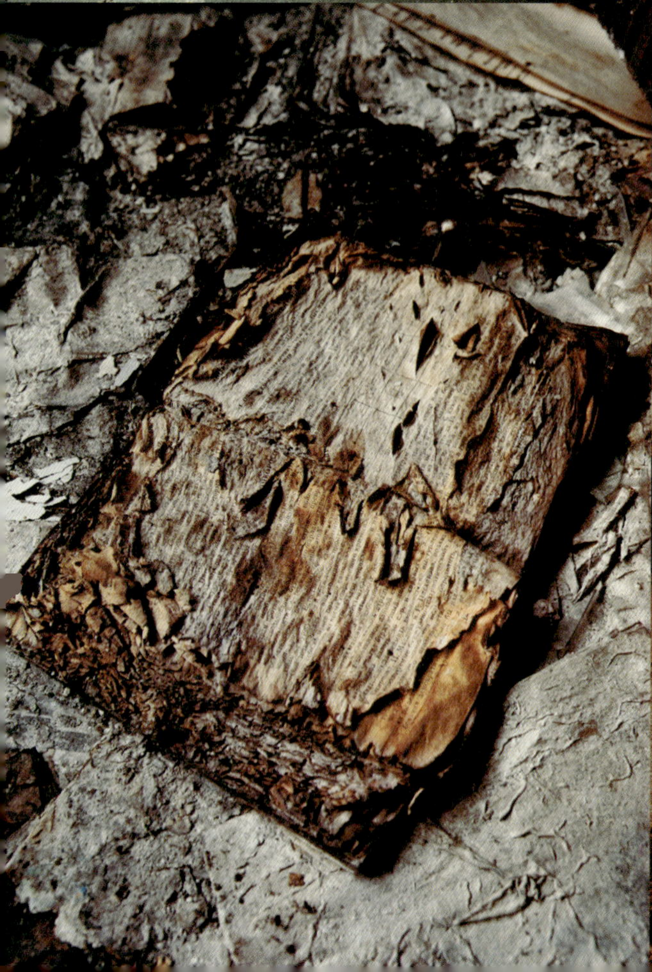

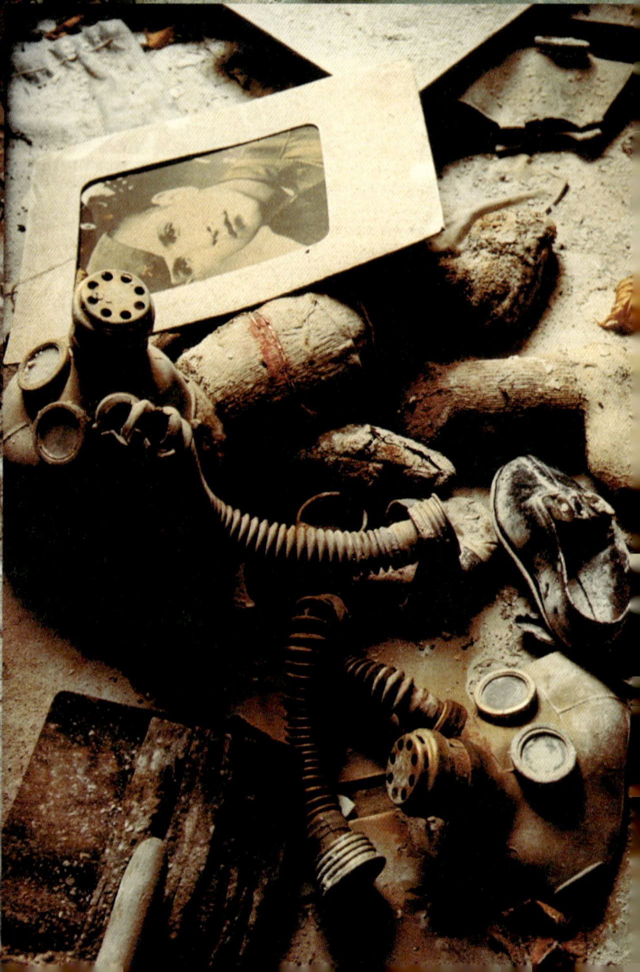

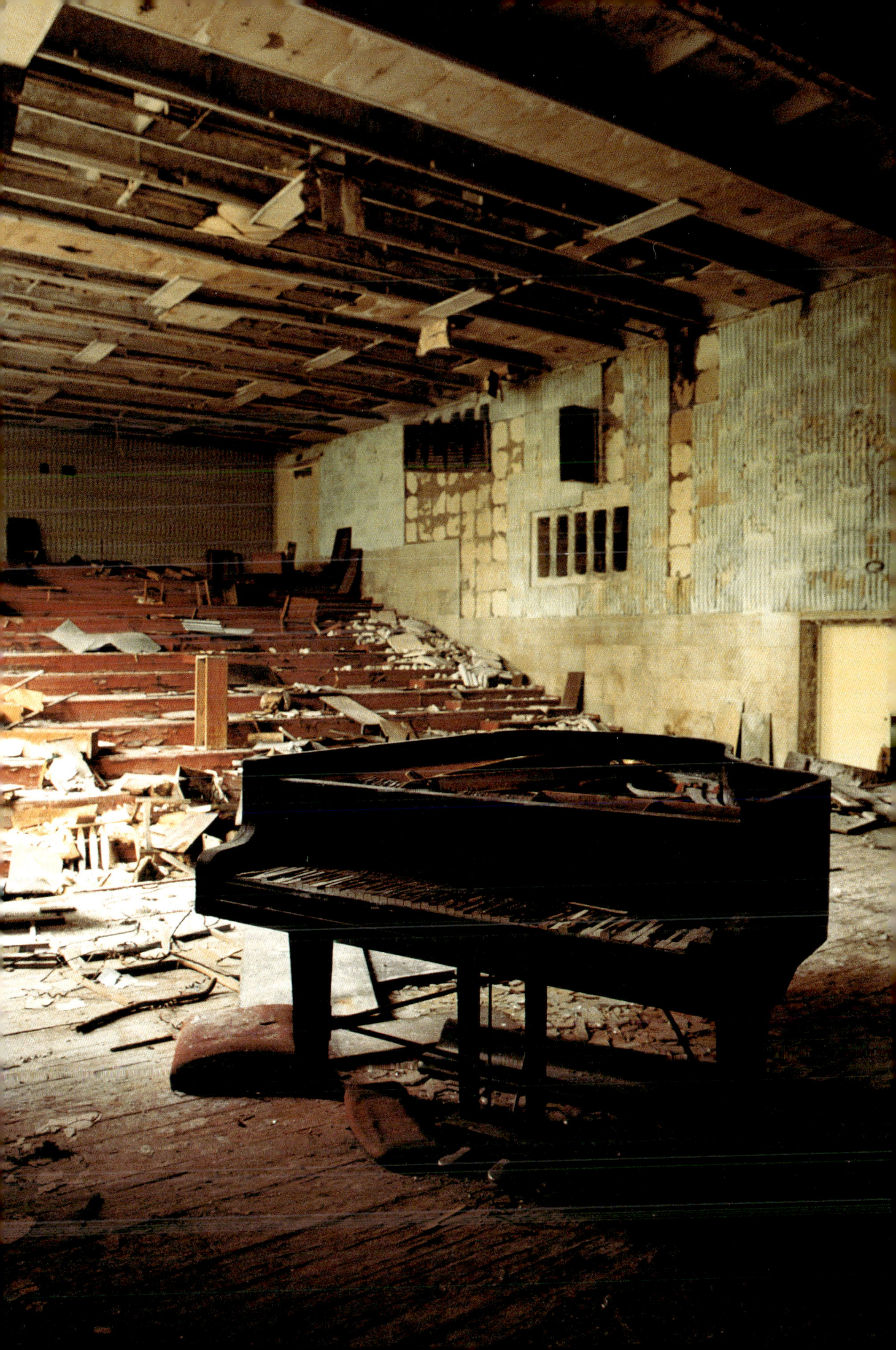

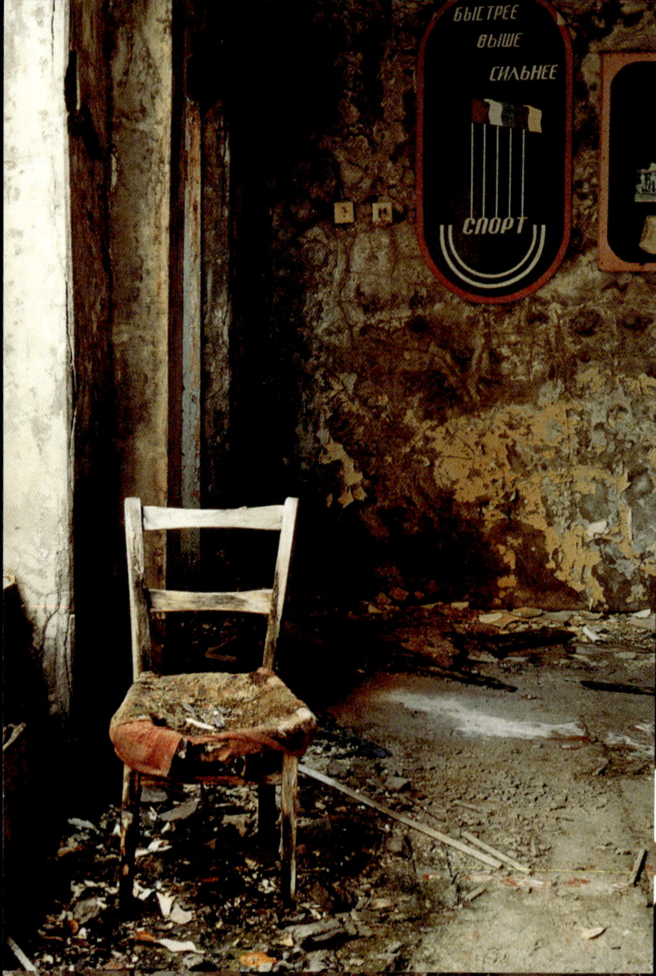
БЫСТРЕЕ
ВЫШЕ
СИЛЬНЕЕ
СПОРТ
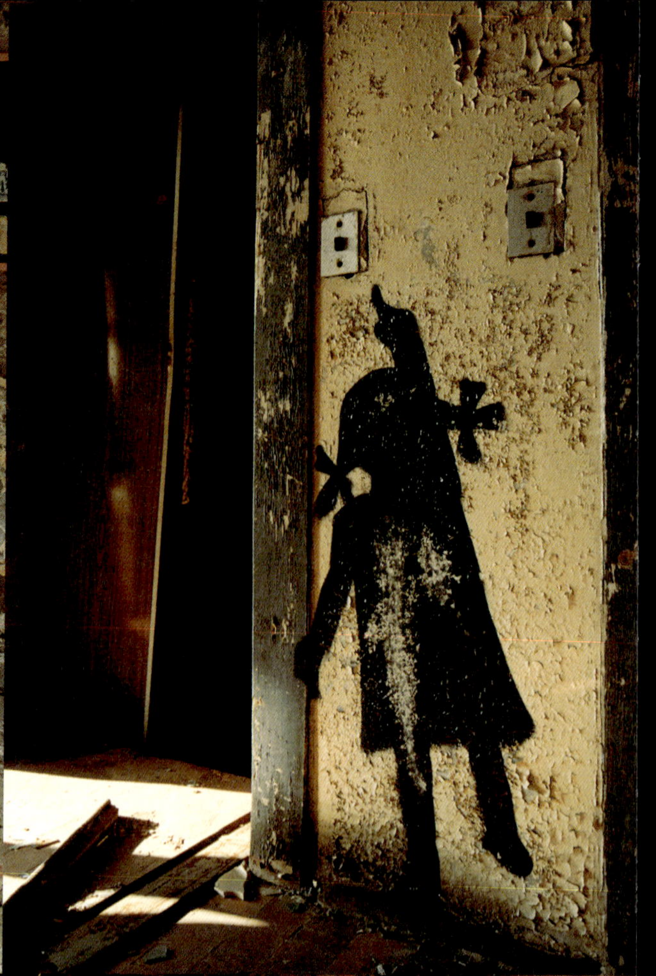
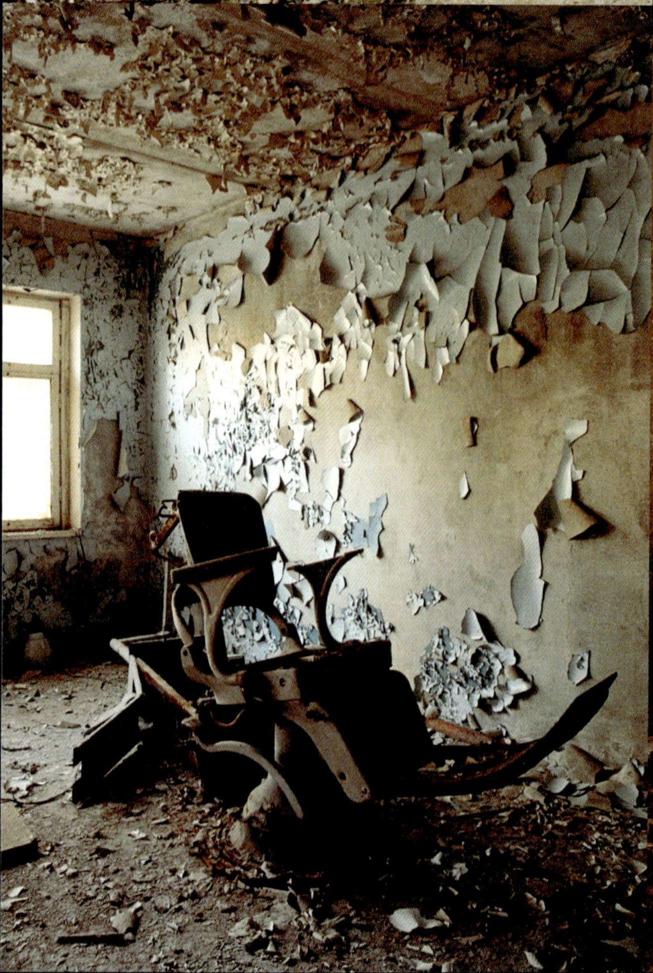
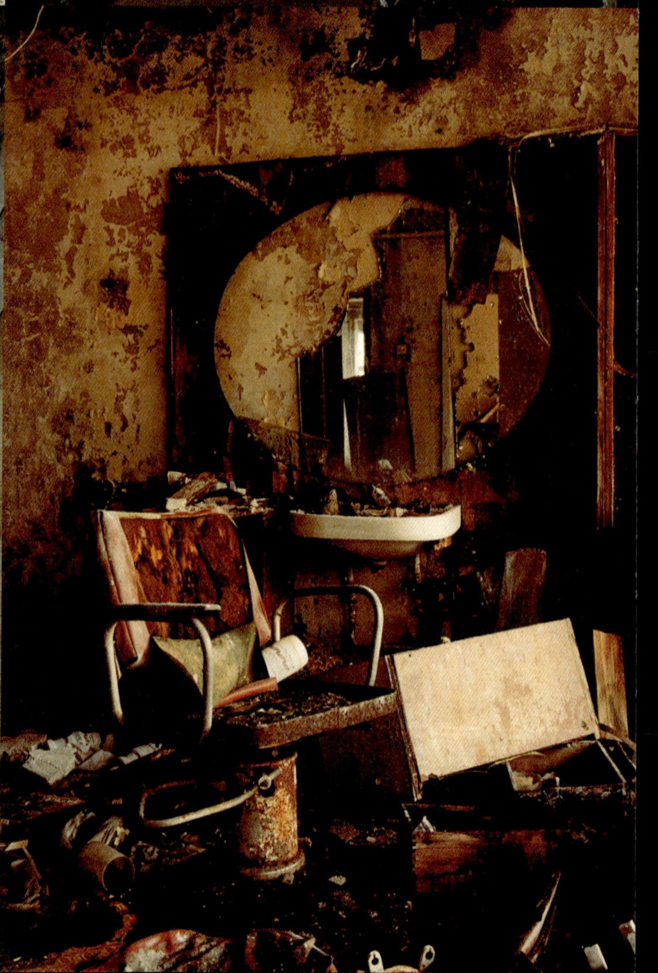

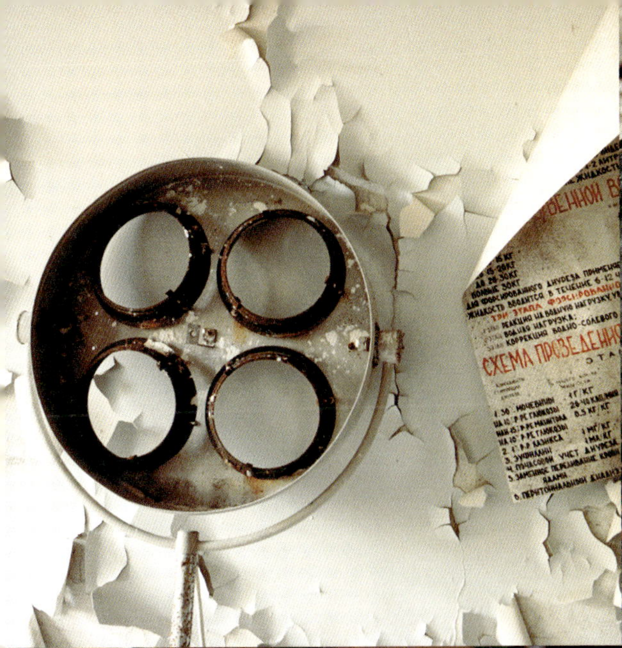
СХЕМА
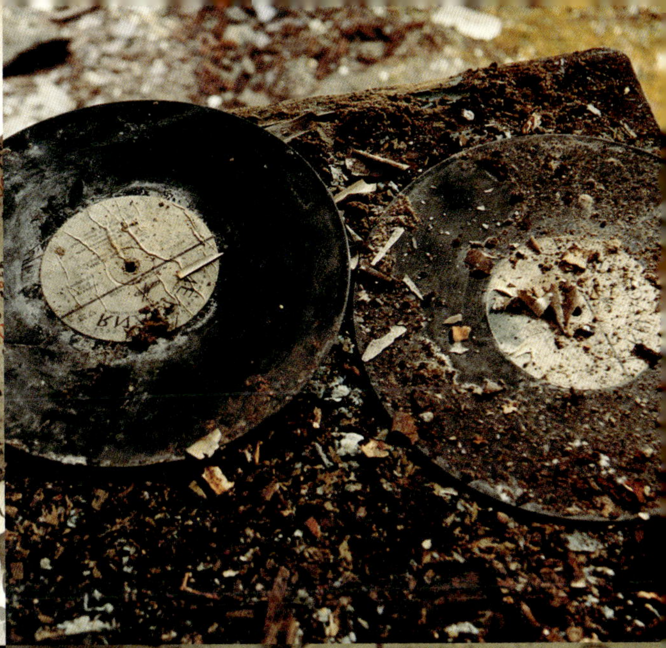
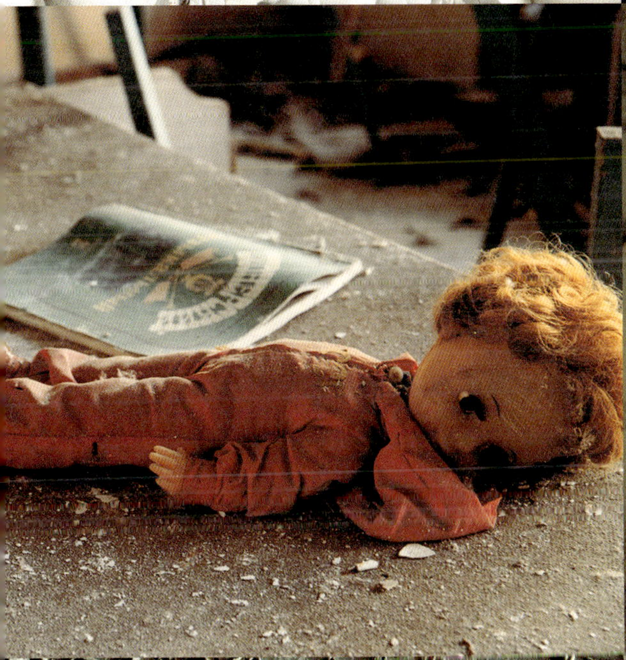
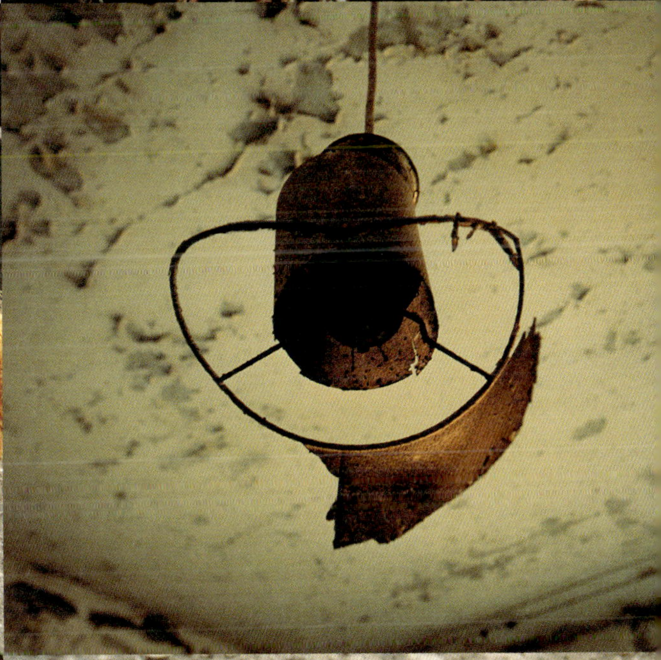
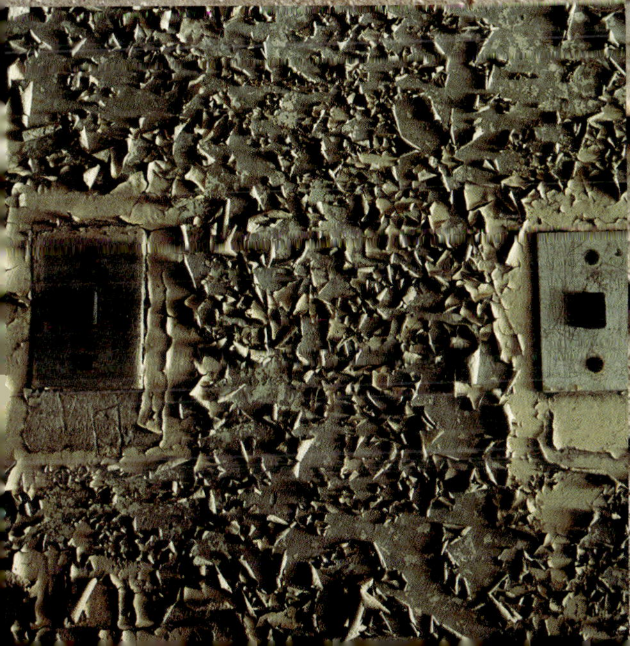
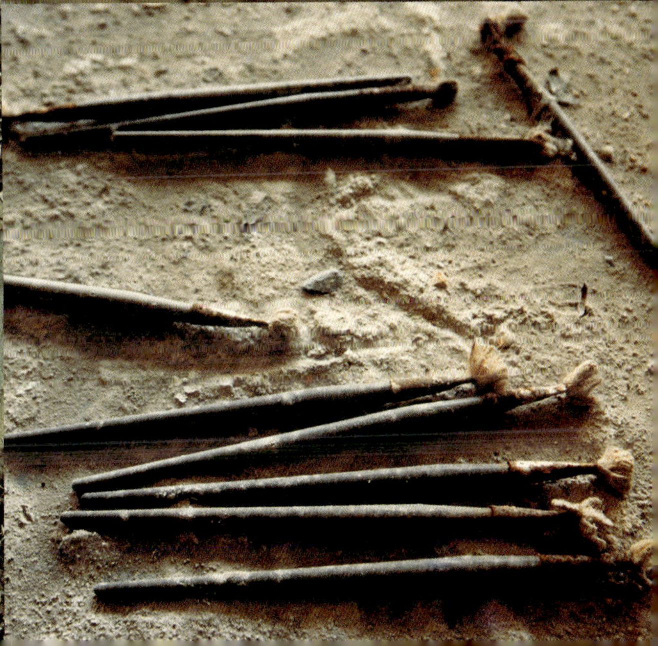

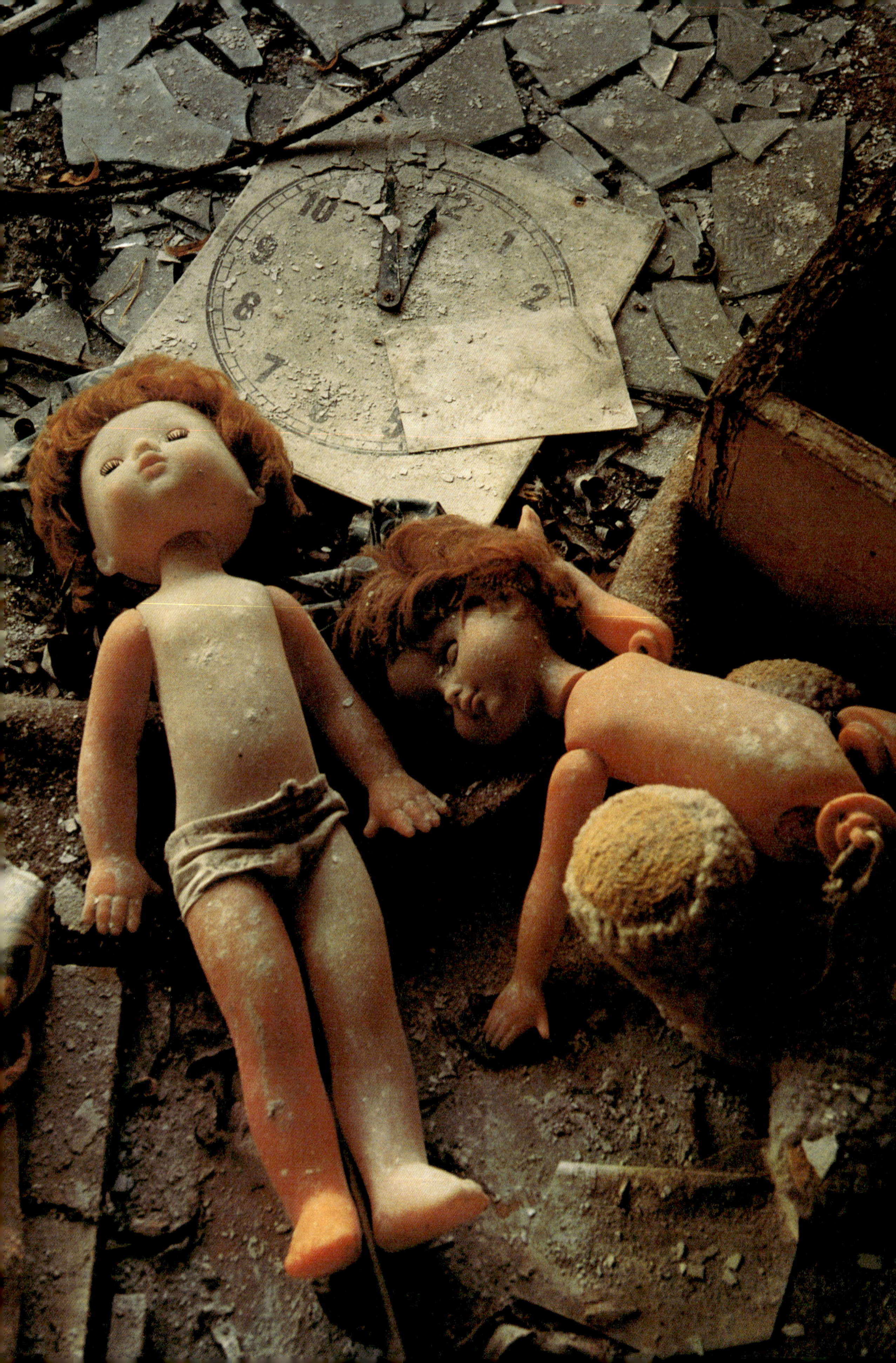

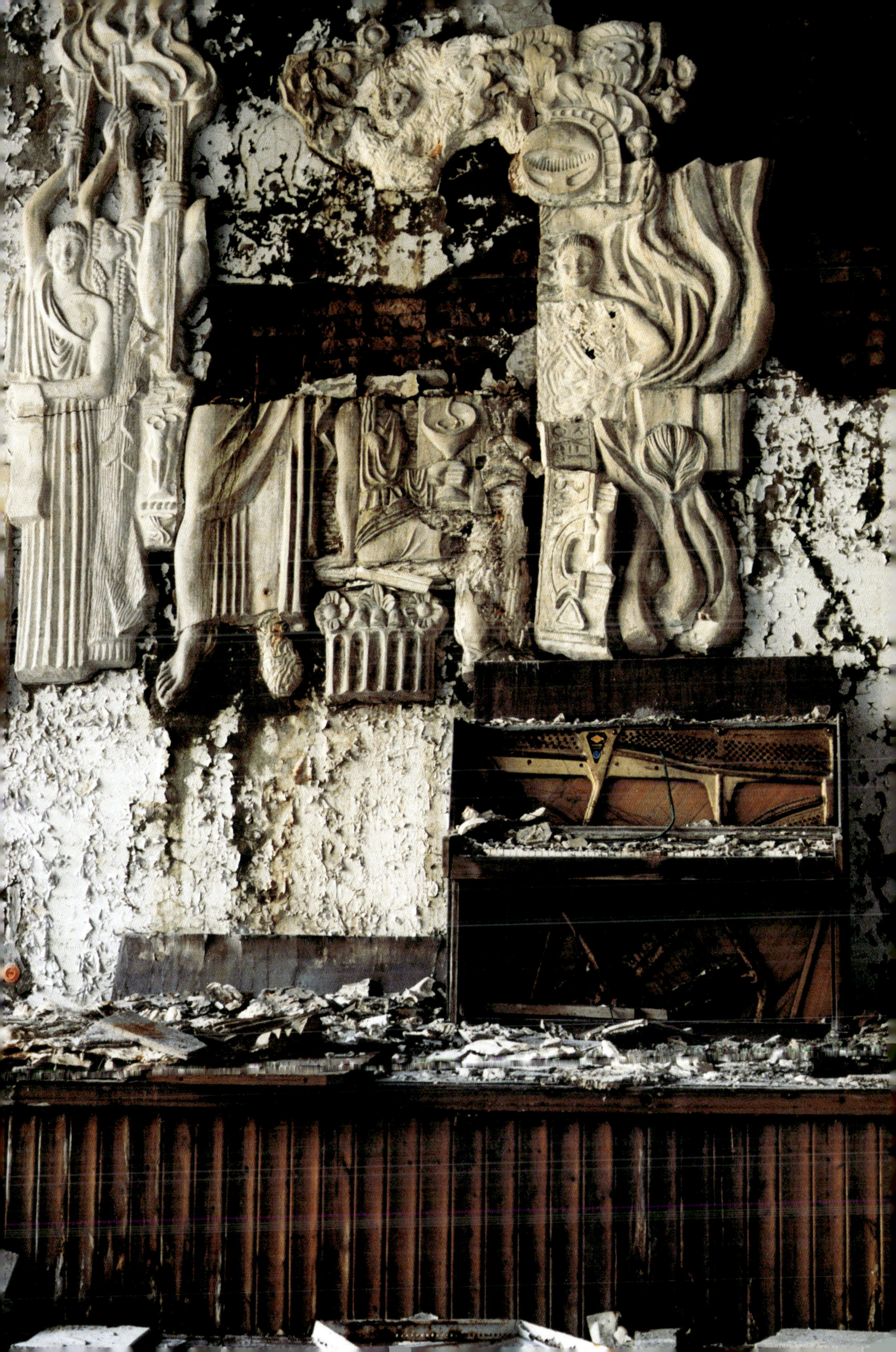

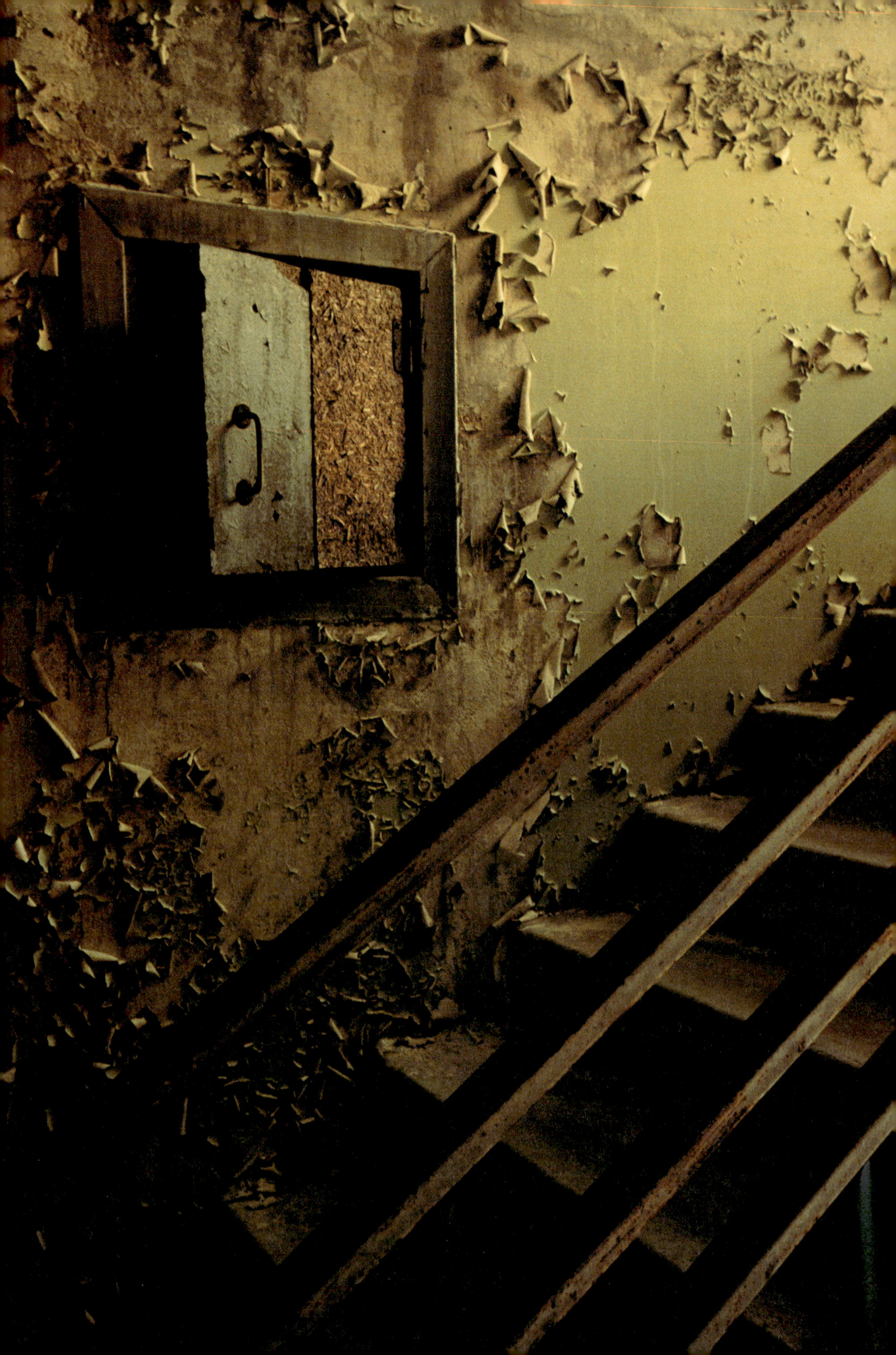

失われたユートピア～プリピァチ　2

「純さん、廃墟の中のものは触ってはいけません。危険です、あぶないです！」

落ちていた人形にそっと手を伸ばしそうになった僕に、通訳のアレクセイがガイガーカウンターを指差して叫んだ。草木も生えぬ大地を想像していた僕は、放射能など素知らぬ顔で黄金色に輝く木々の逞しさを目の当たりにし、この地が放射能汚染されていることを忘れかけていた。ガイガーカウンターは突如予期せぬ場所で鳴り始める。そのけたたましい電子音に一瞬身がすくむのだが、この地を旅するための重要なお守りであることには間違いない。

ウクライナ人スタッフと僕は、林立する建物の中を歩き続けた。主を失って22年も経った建築物は、大部分が苔むしひび割れ、瓦礫の山が散在する予想通りの終末の光景である。建物の壁はほとんどが地のコンクリートに厚くペンキを塗っただけなのか、ペンキの層が薄く不規則にはがれて仄かな明かりに照らされている。まさに廃墟に咲く徒花のようだ。

目に入るもの大半がソ連という国家が目指した実質主義、効率主義を具現しているようである。画一的な建物はもちろんのこと、打ち捨てられた家具調度品に至ってはベニヤ板の工作のようである。だがそんな殺風景な廃墟にあって唯一心が和むのが、所々で出会うピアノの残骸だ。ピアノに少し心得のあるガイド、サーシャは放射能レベルの確認をした後、ほこりがたまった鍵盤に指を滑らせる。調律の狂った不思議な調べが殺風景なコンクリートの箱に響き渡り一瞬だけ命を吹き込んでいるかのようである。

ここに住んでいたのは、まぎれもなく音楽を聴くふつうの人間だったのだ。

Lost Utooia Pripiati 2

"Don't touch it, Jun-san. It's dangerous, dangerous!"
The interpreter, Alexei shouted when I was going to reach out for a doll on the ground. I believed that no plant would grow on the polluted ground by radioactivity. But seeing sturdy golden trees neglecting fearful influence of radioactivity, I was apt to forget the polluted soil in the area. The Geiger counter suddenly began to tell danger. We always coward at the terrible sound of this counter, but this kind of apparatus is certainly the guardian for the travelers in this area.
A Ukrainian staff and I were walking on through a forest of buildings. A building which lost its residents ?? years ago and covered with mossy cracks showed an anticipated result of the end of the world. Almost all the walls of the building were only thickly painted, and so the layer of the thick paint came off the surface in places and reflected bright light. It was like an abortive flower.
Everything we could see appeared to have been embodied the pragmatism and efficiency which the Soviet Union aimed at. Not to say of the standardized building, deserted furniture was like the works of plywood. But in such bleak ruins, the only thing which comforted us was the wrecked pianos we found here and there. A guide, Sasha who had an interest in piano playing, after acknowledged the level of radioactivity, played the piano which was covered with dust. The strange tune from the out of ordered piano echoed in the bleak concrete box and seemed to have given a life to the atmosphere.
The residents who lived in this house must have been an ordinary people.

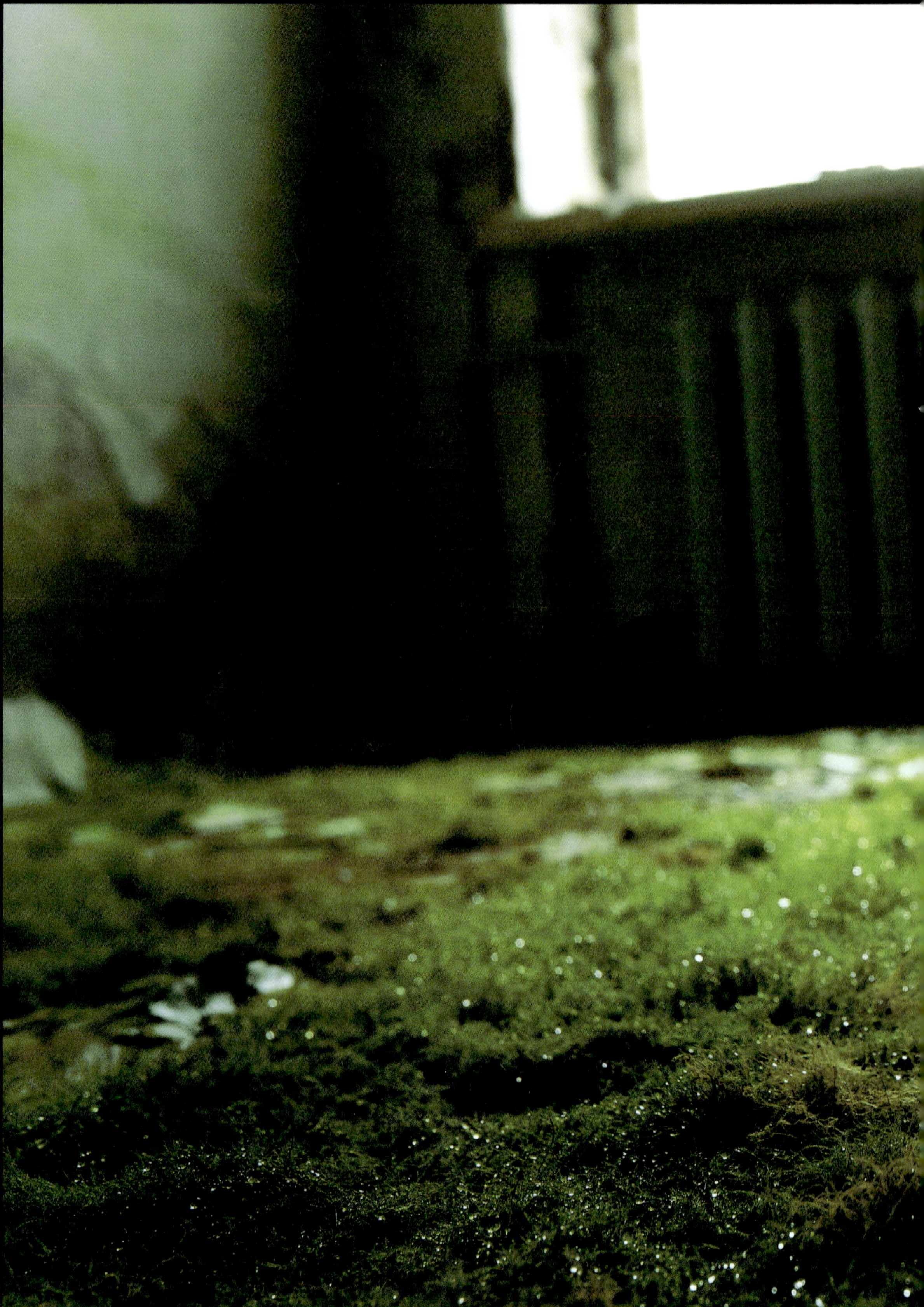

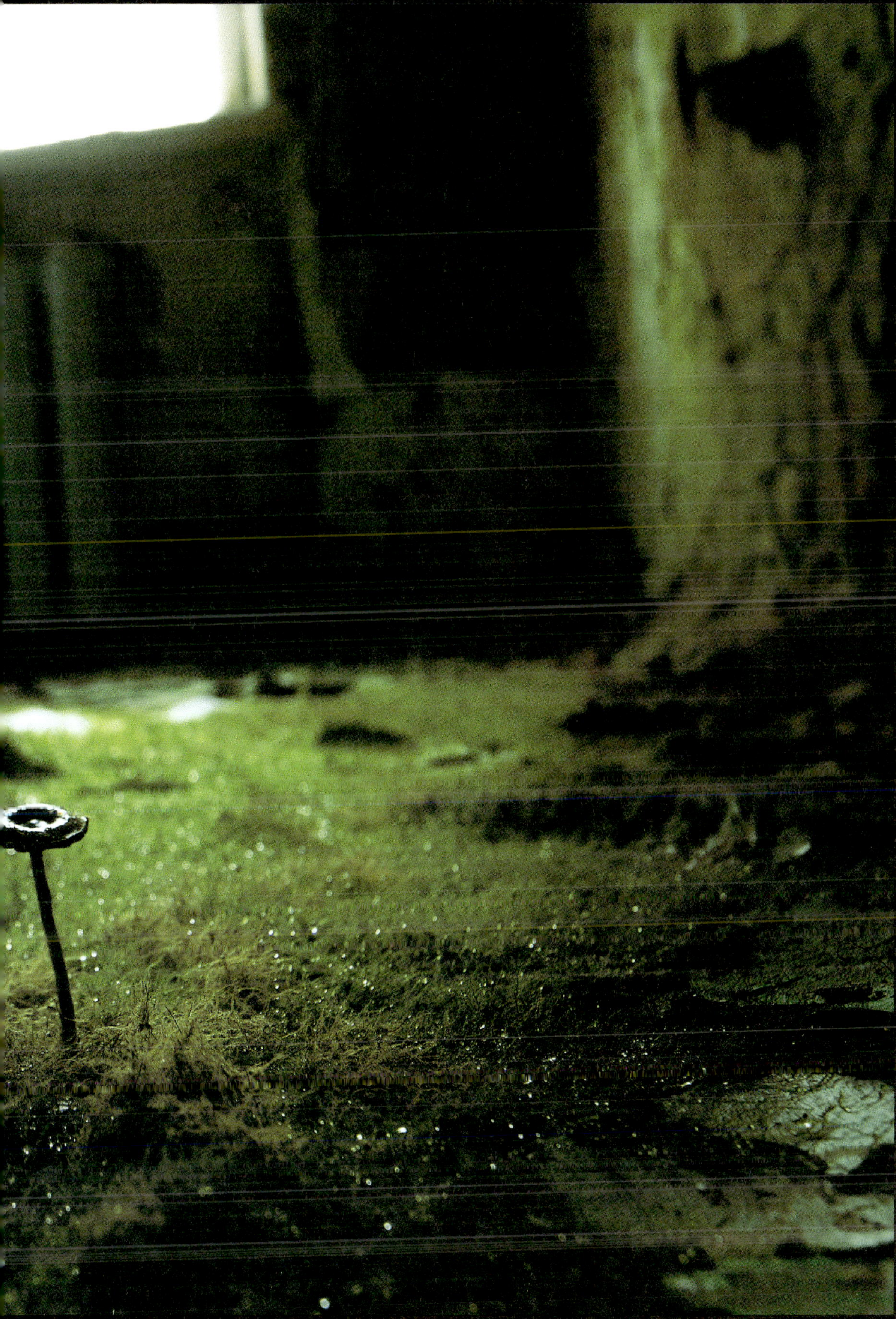

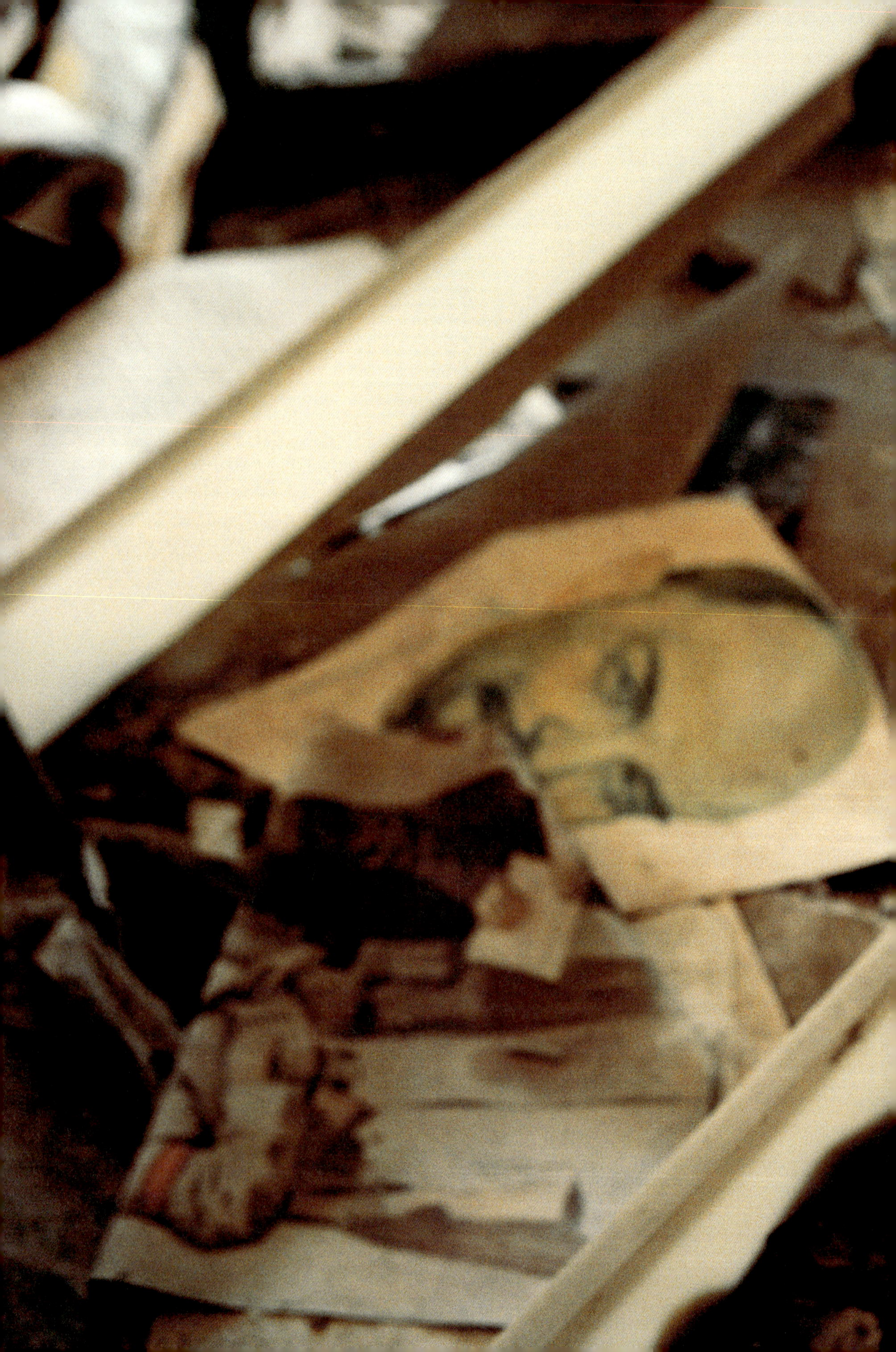

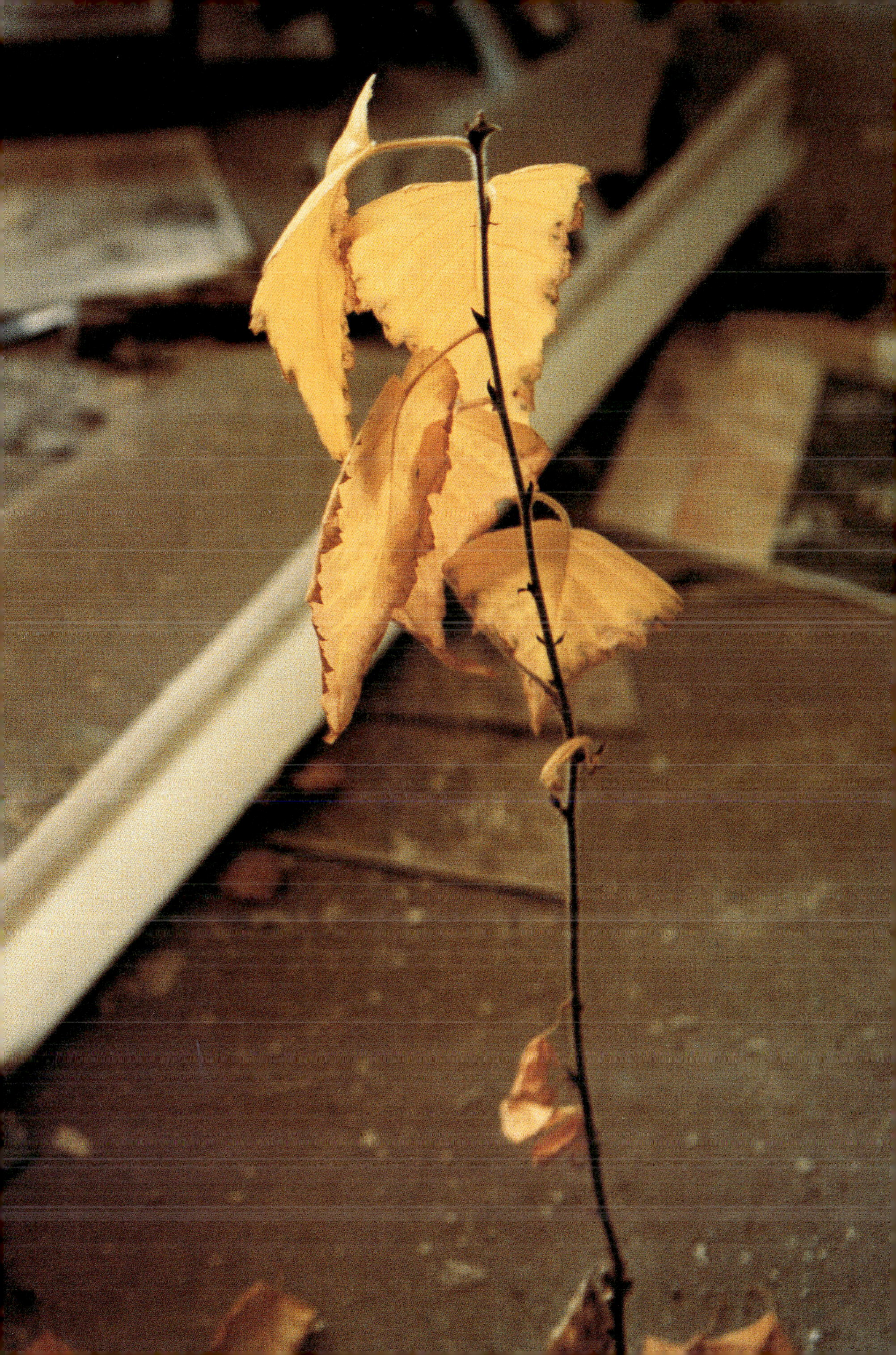

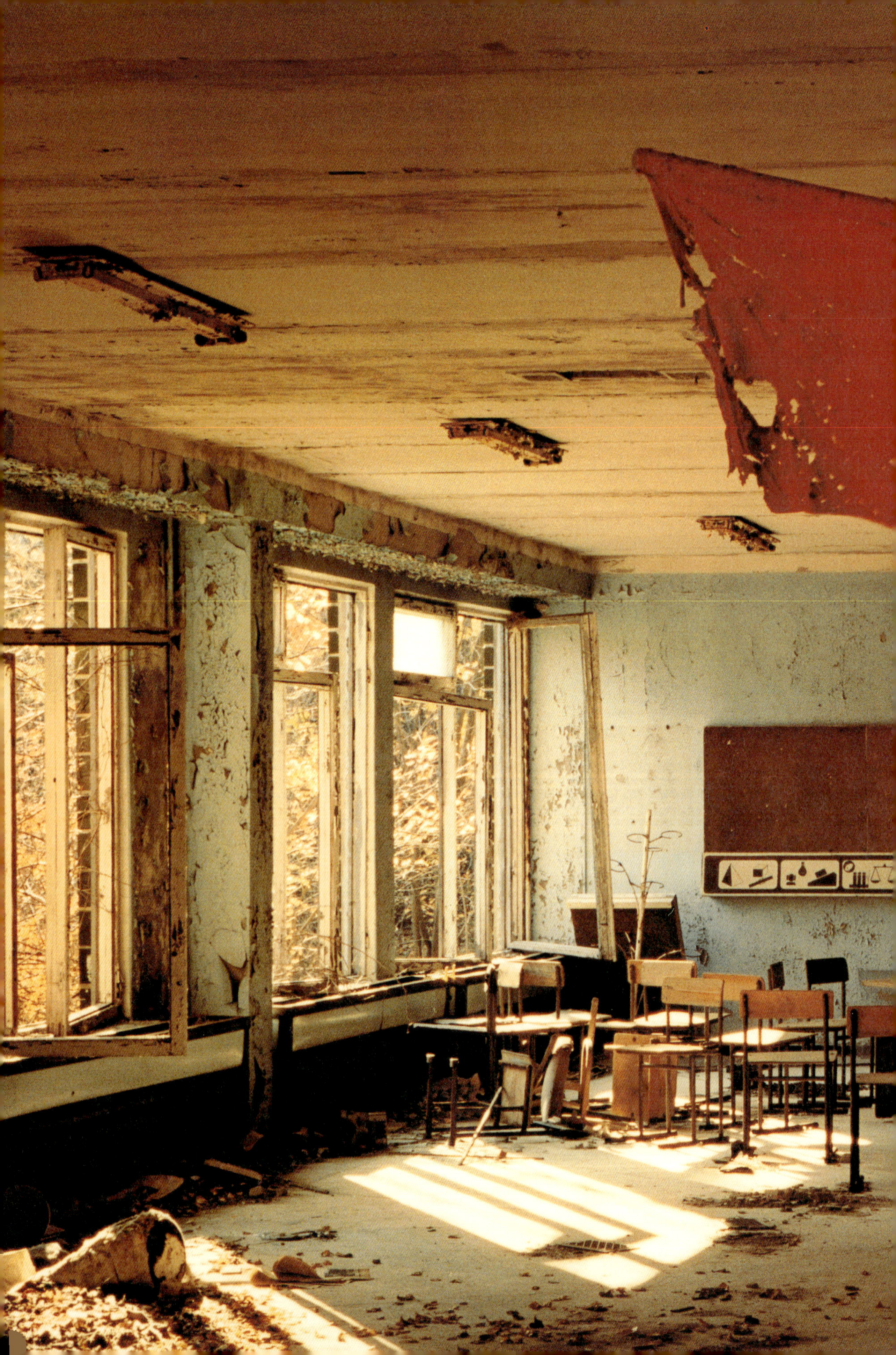

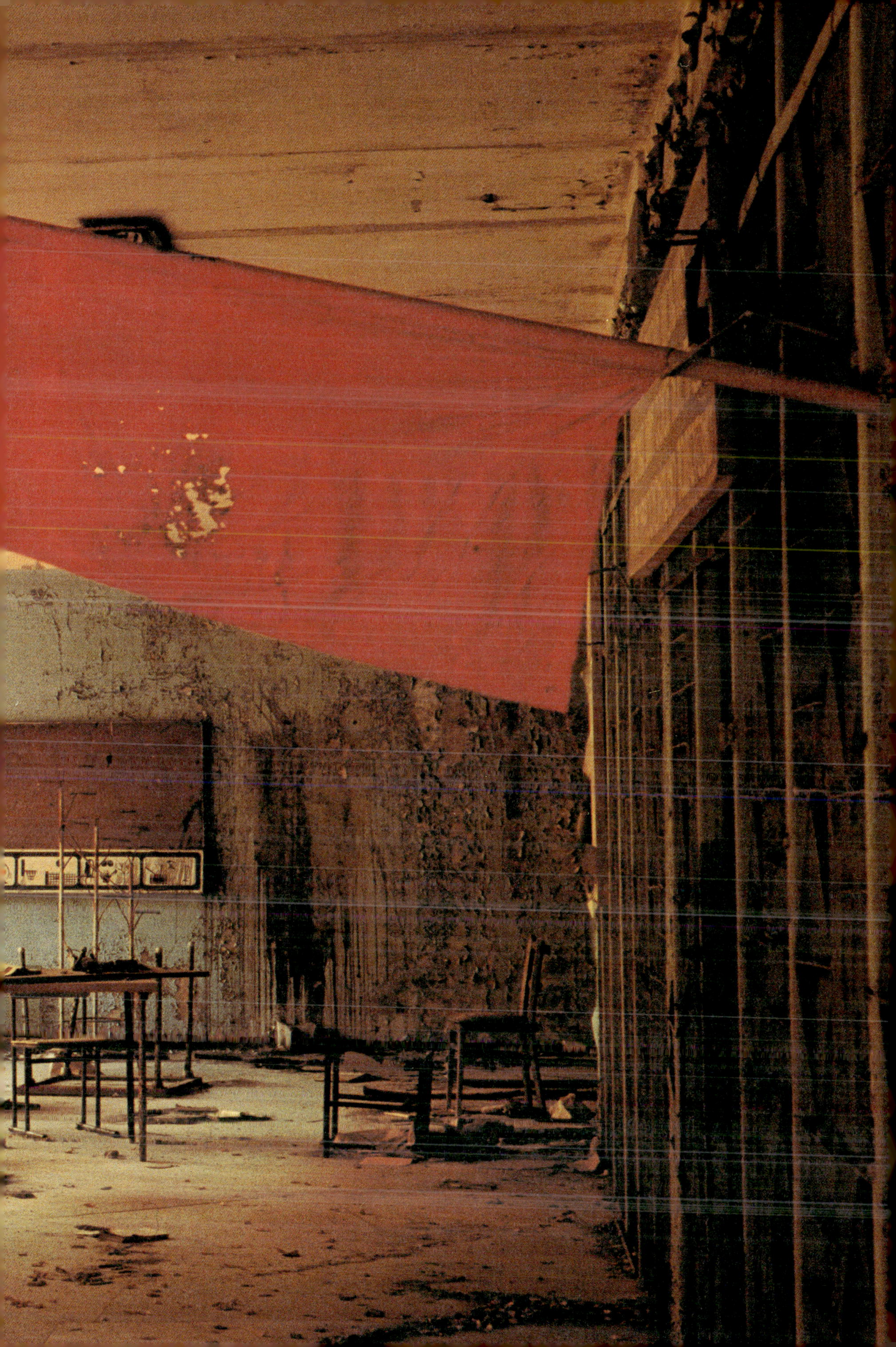

1945-1985

БЕССМЕРТЕН
ТВОЙ
ПОДВИГ,
СОЛДАТ!

РЯДОВОЙ
ГОЛОСИНА А.И.
ЛЕЙТЕНАНТ
КАЛАЧЕВ А.В.
ГВ ЛЕЙТЕНАНТ
КАЛИНИЧЕНКО Г.М.
СТ СЕРЖАНТ
РАЙДУПКИН С.П.

БЕСПРИМЕРНЫЙ
ГЕРОИЗМ
ПАРТИЗАН-
ВЕЛИКИЙ
ВКЛАД
В ПОБЕДУ

АТОМ
РАБОТАТЬ
МИР НА КОММУНИЗМ

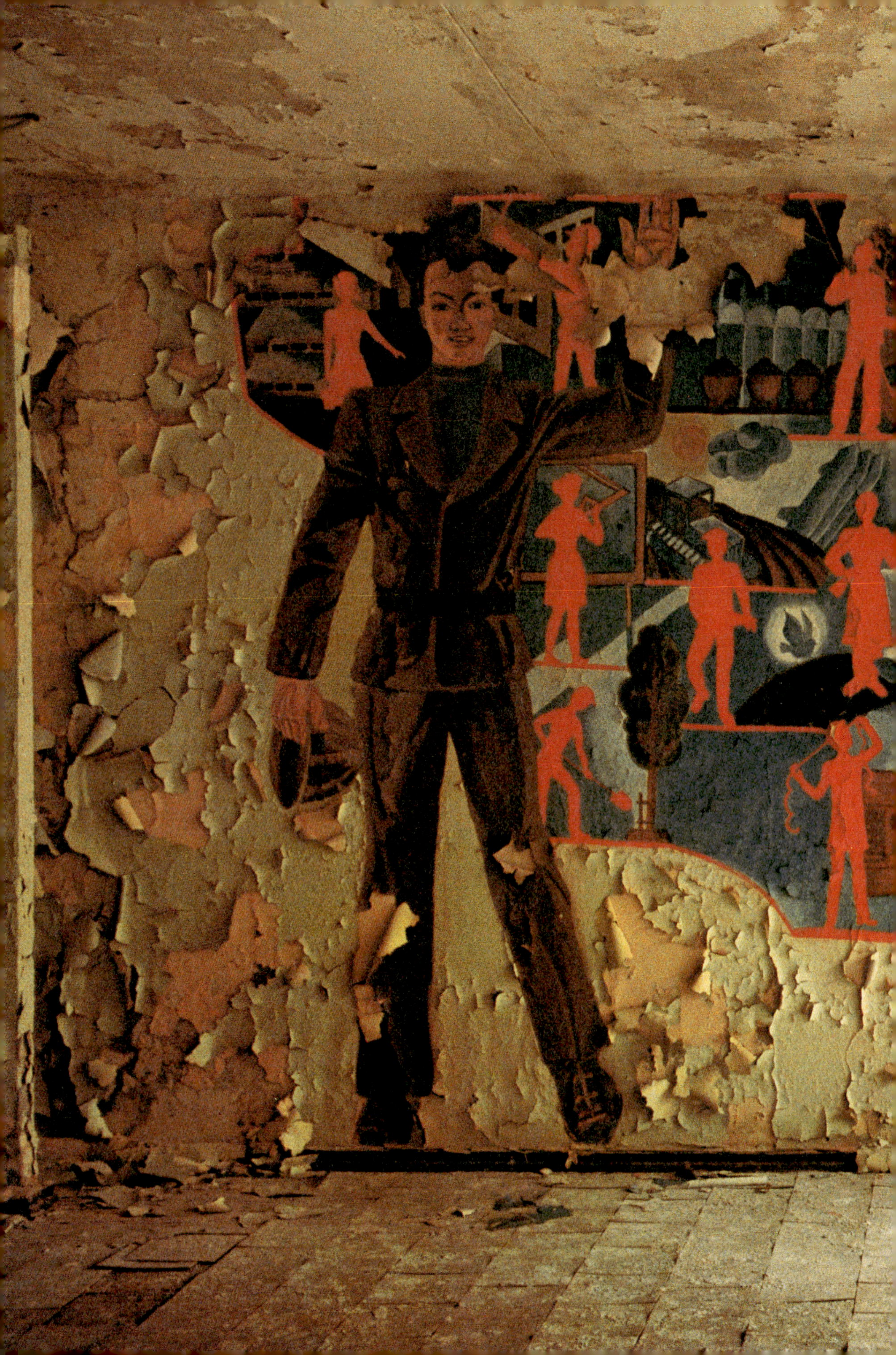

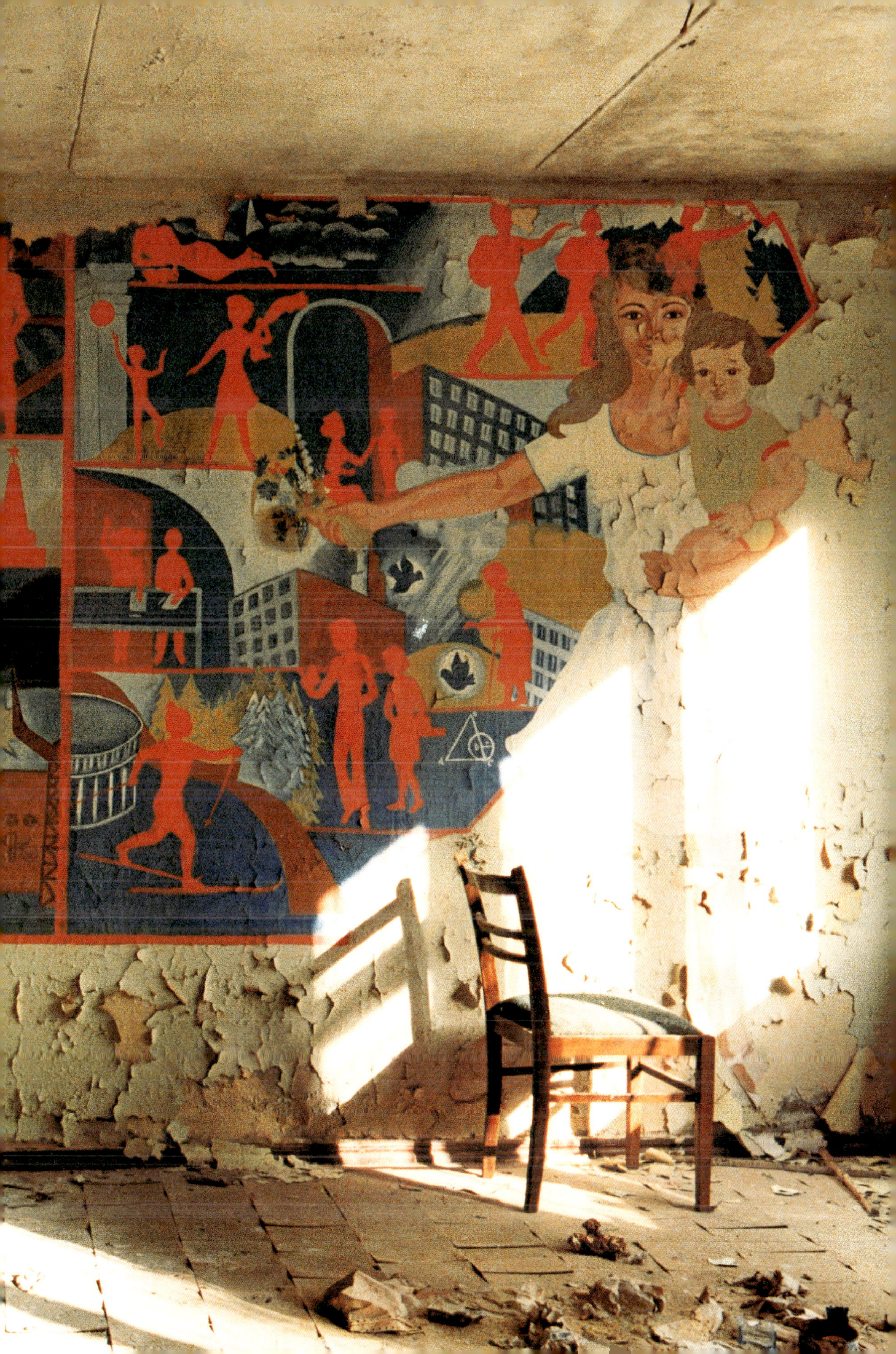

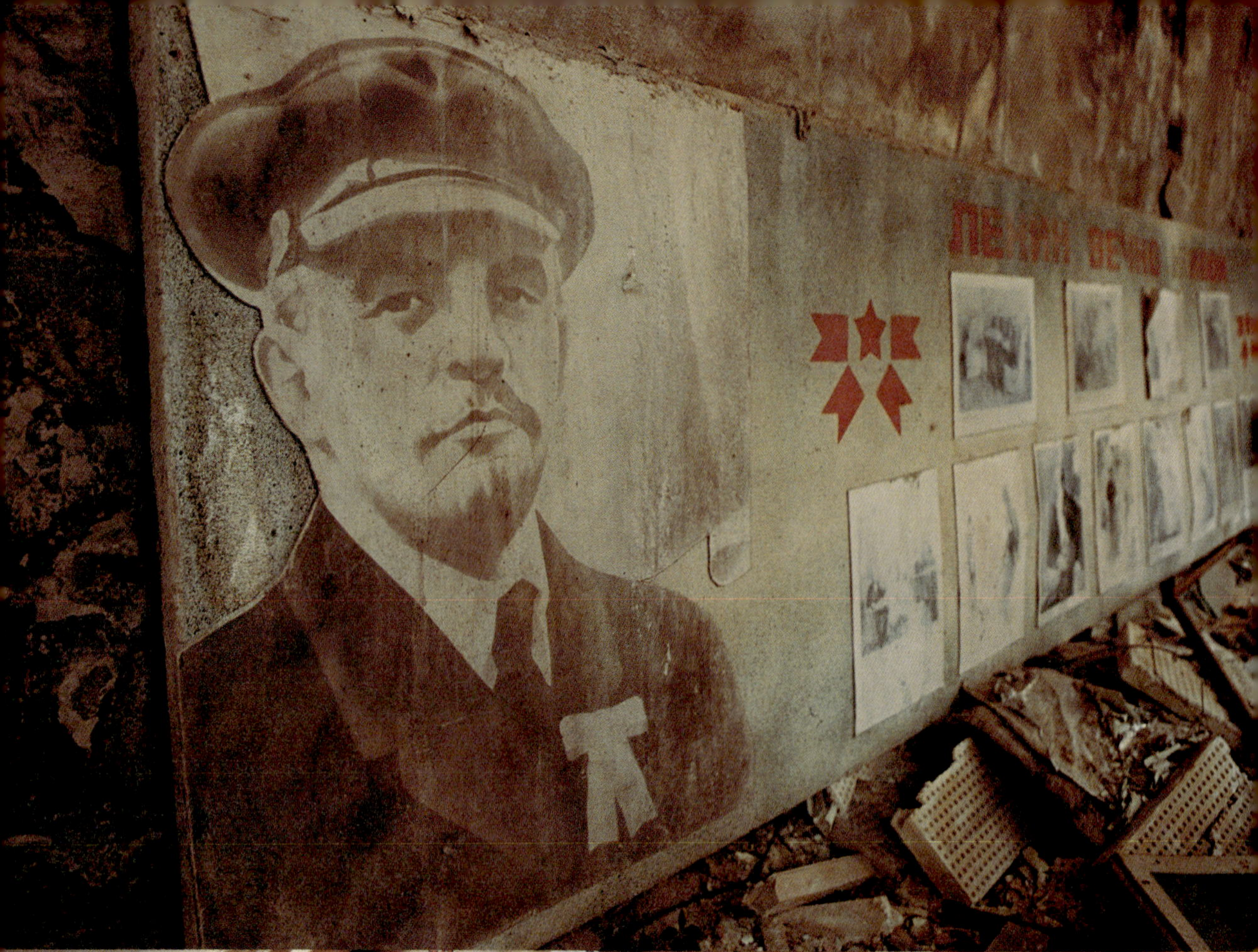

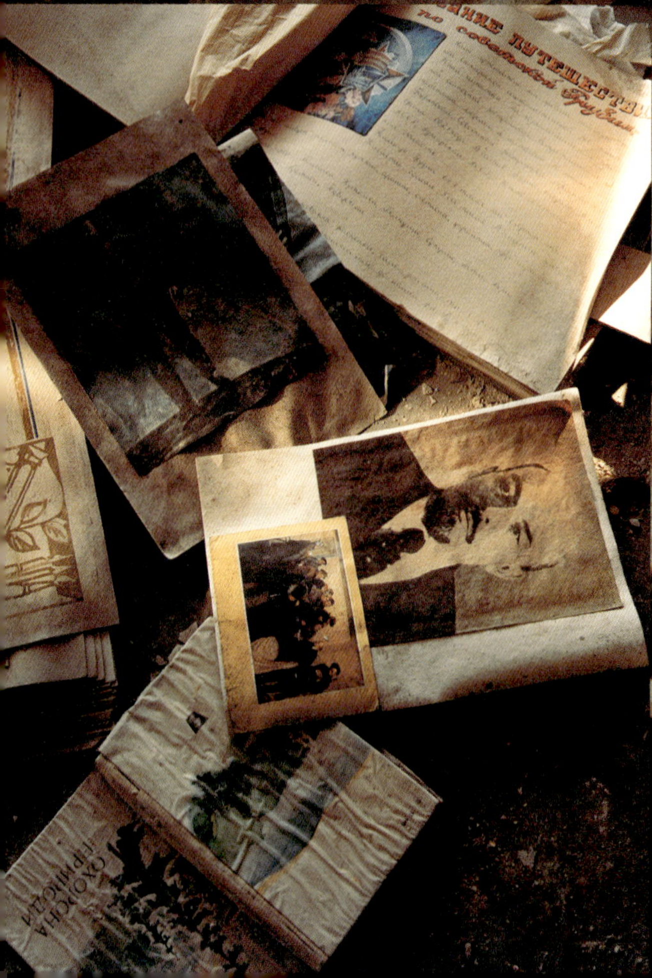

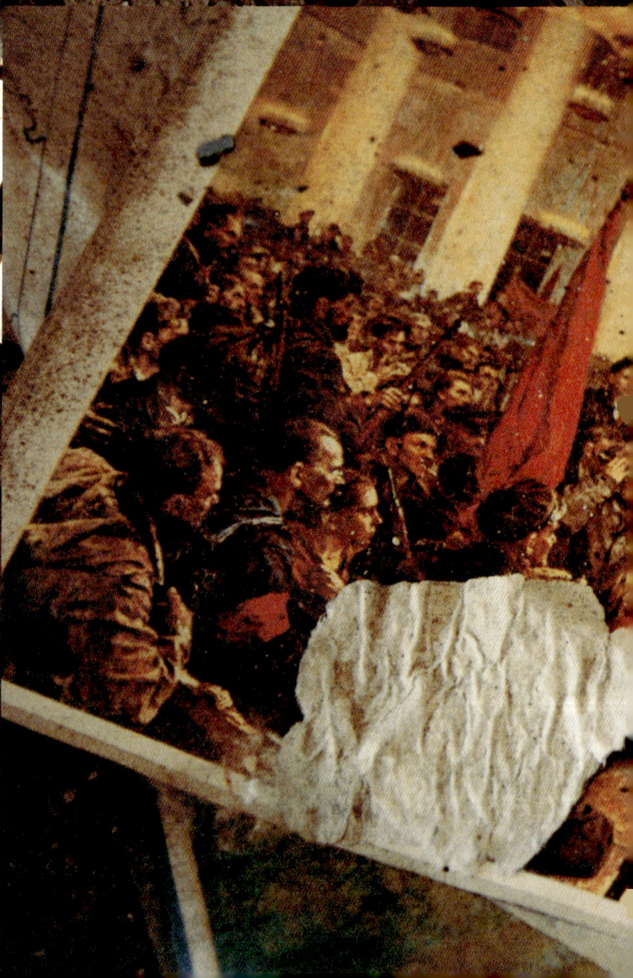

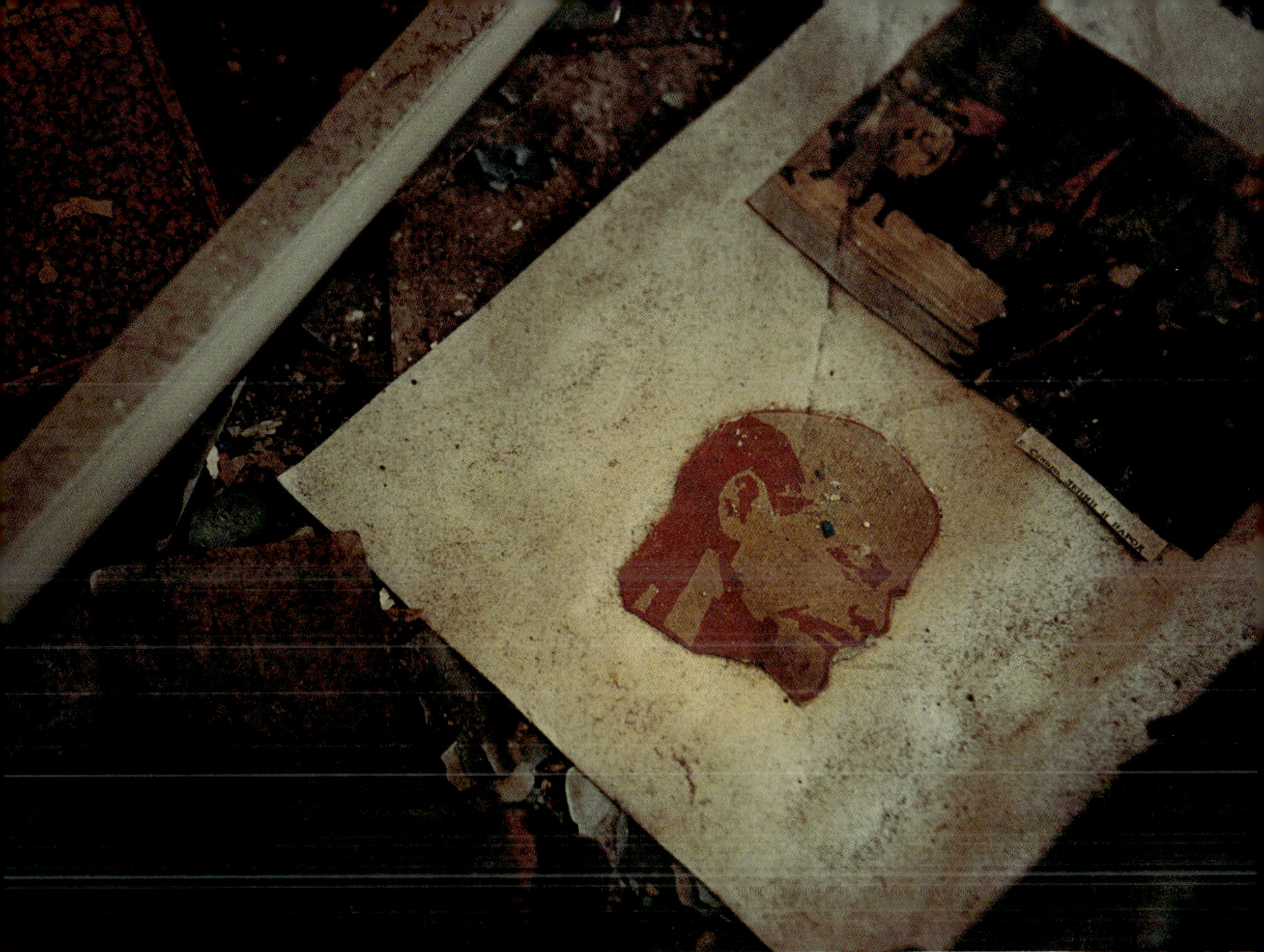

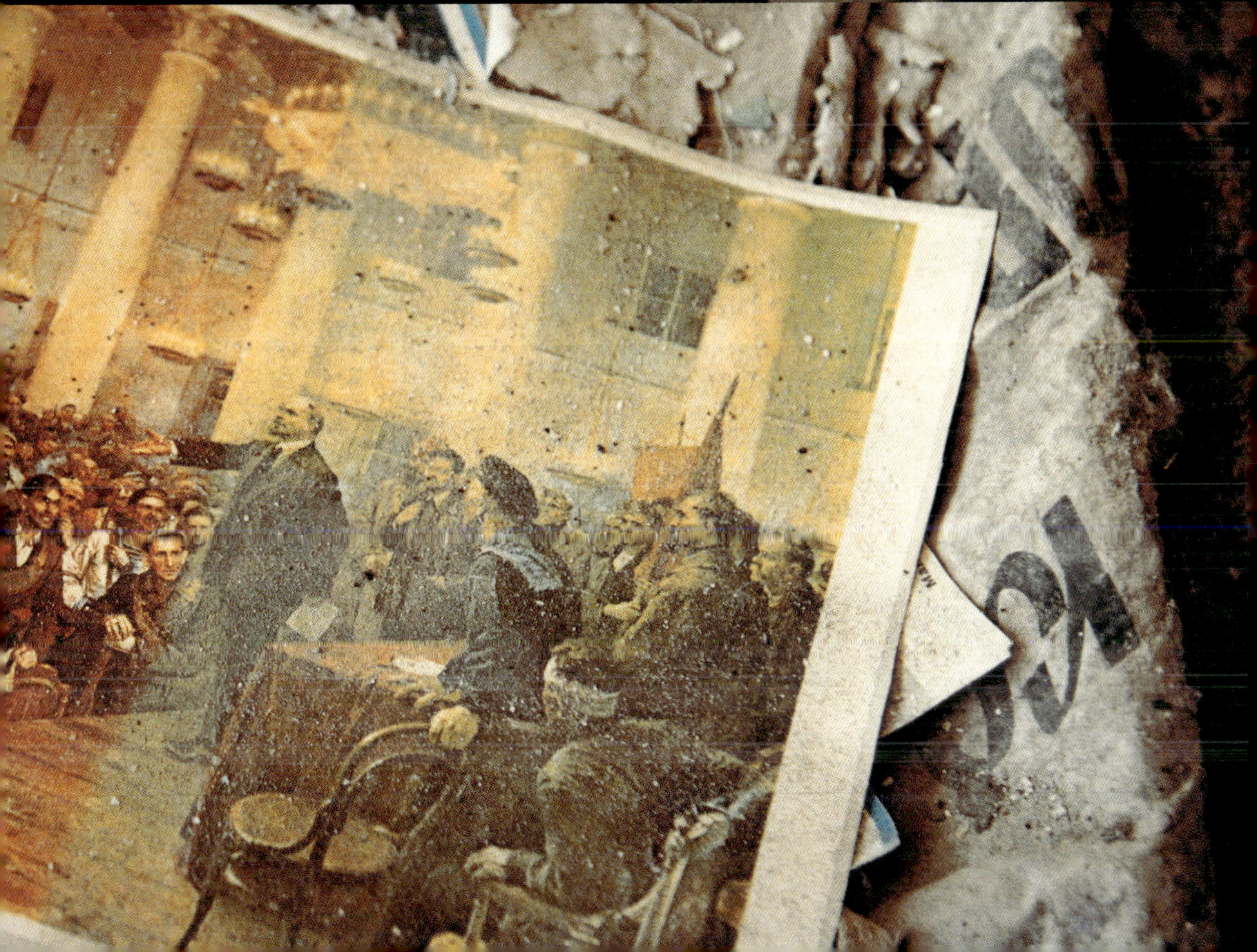

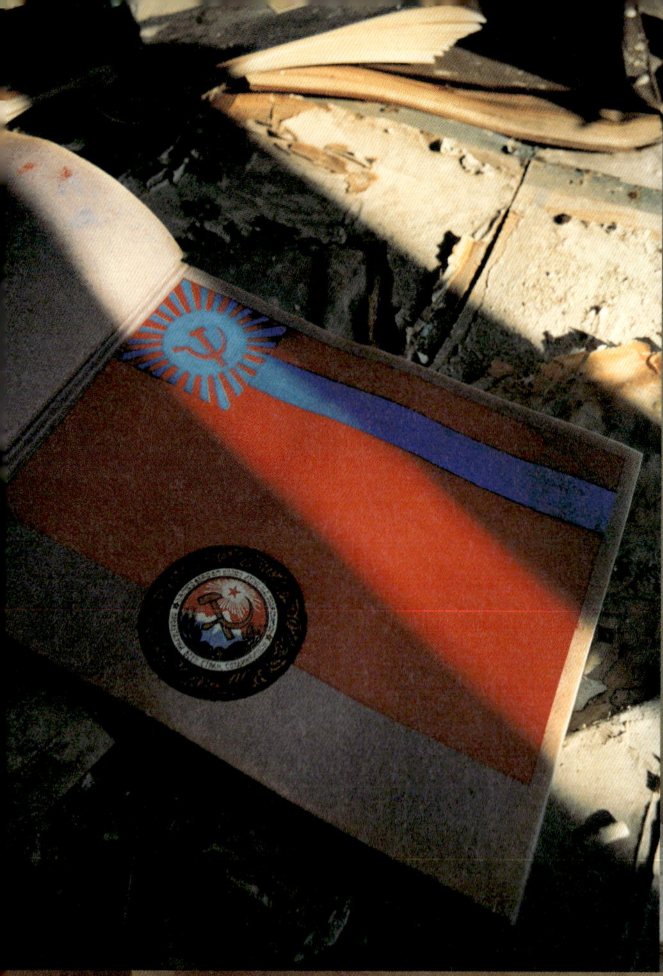

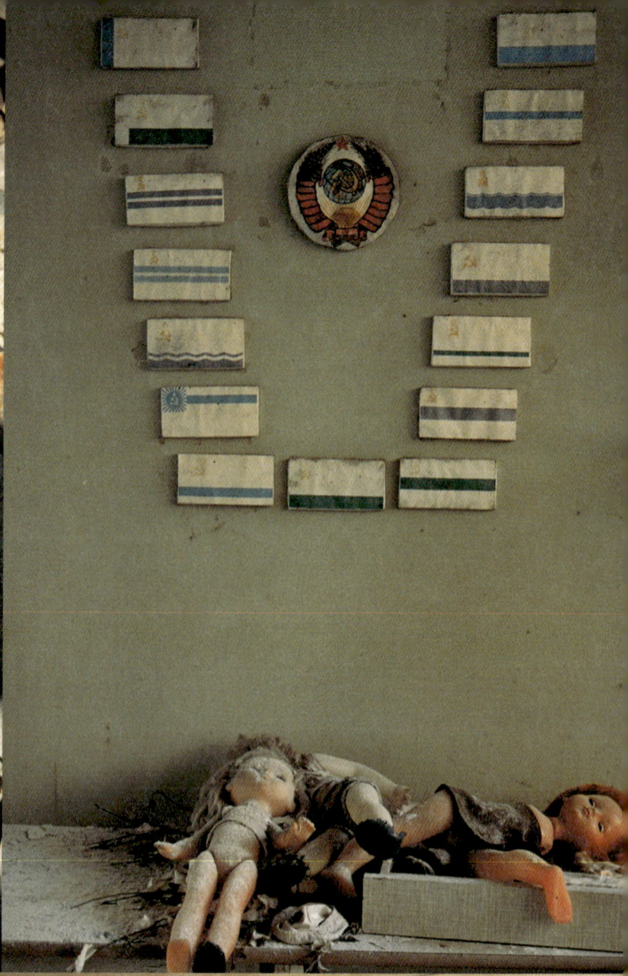

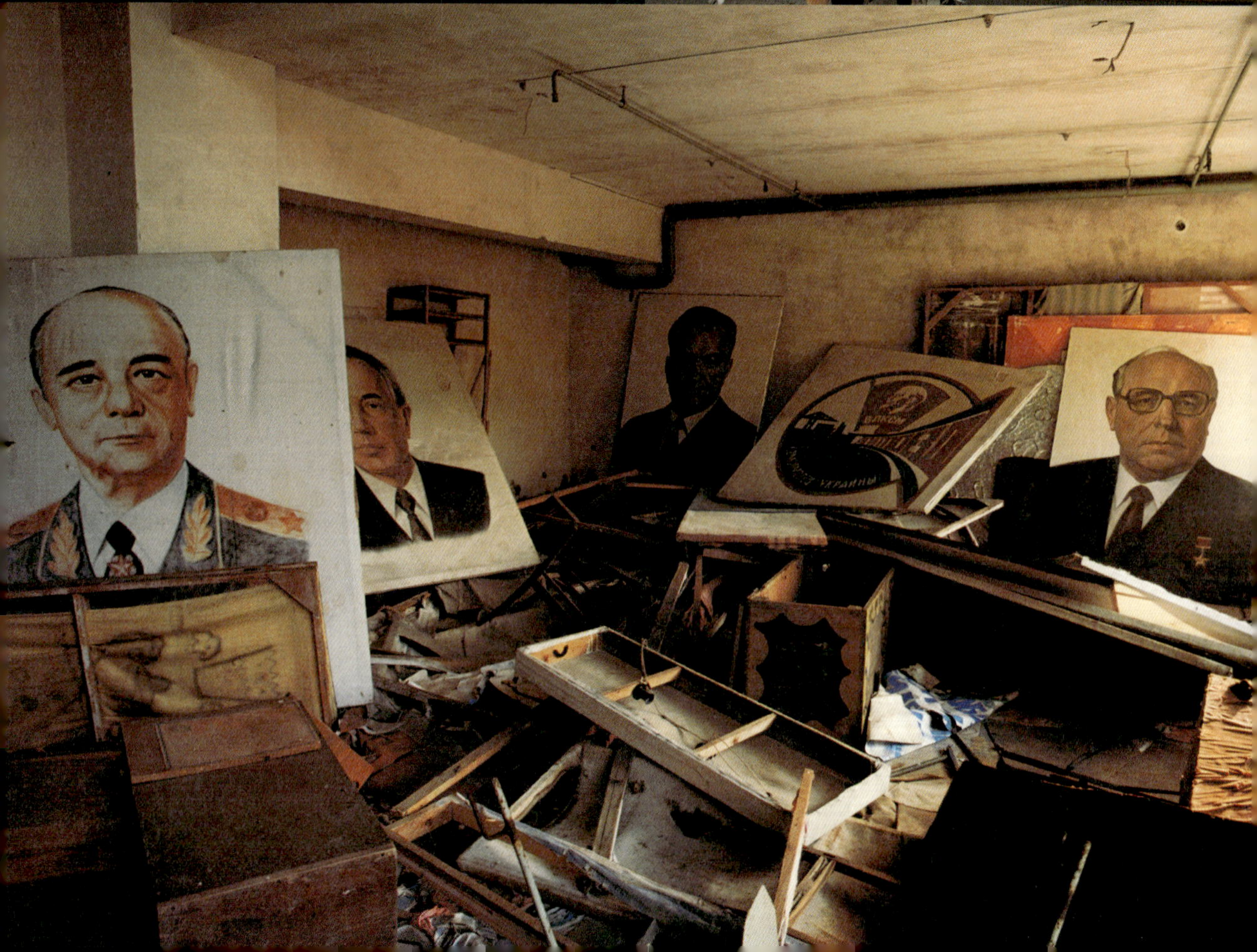

Тт Тт Тт Тт Тт
Дон, Кама, Днепр, Двина.
Лена, Амур — реки.

ГРАЖДАНСКАЯ ОБОРОНА —
ДЕЛО ВСЕНАРОДНОЕ

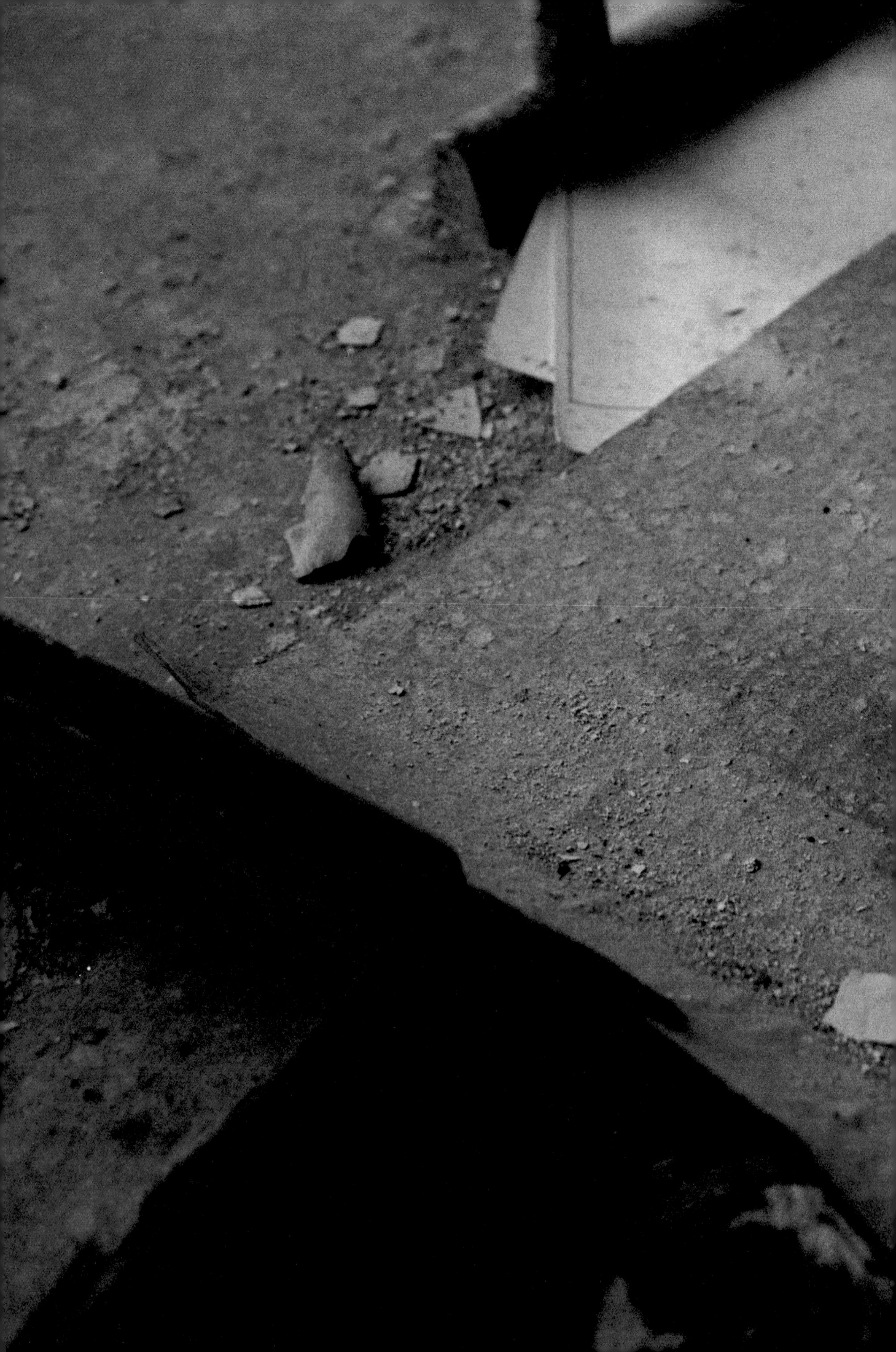

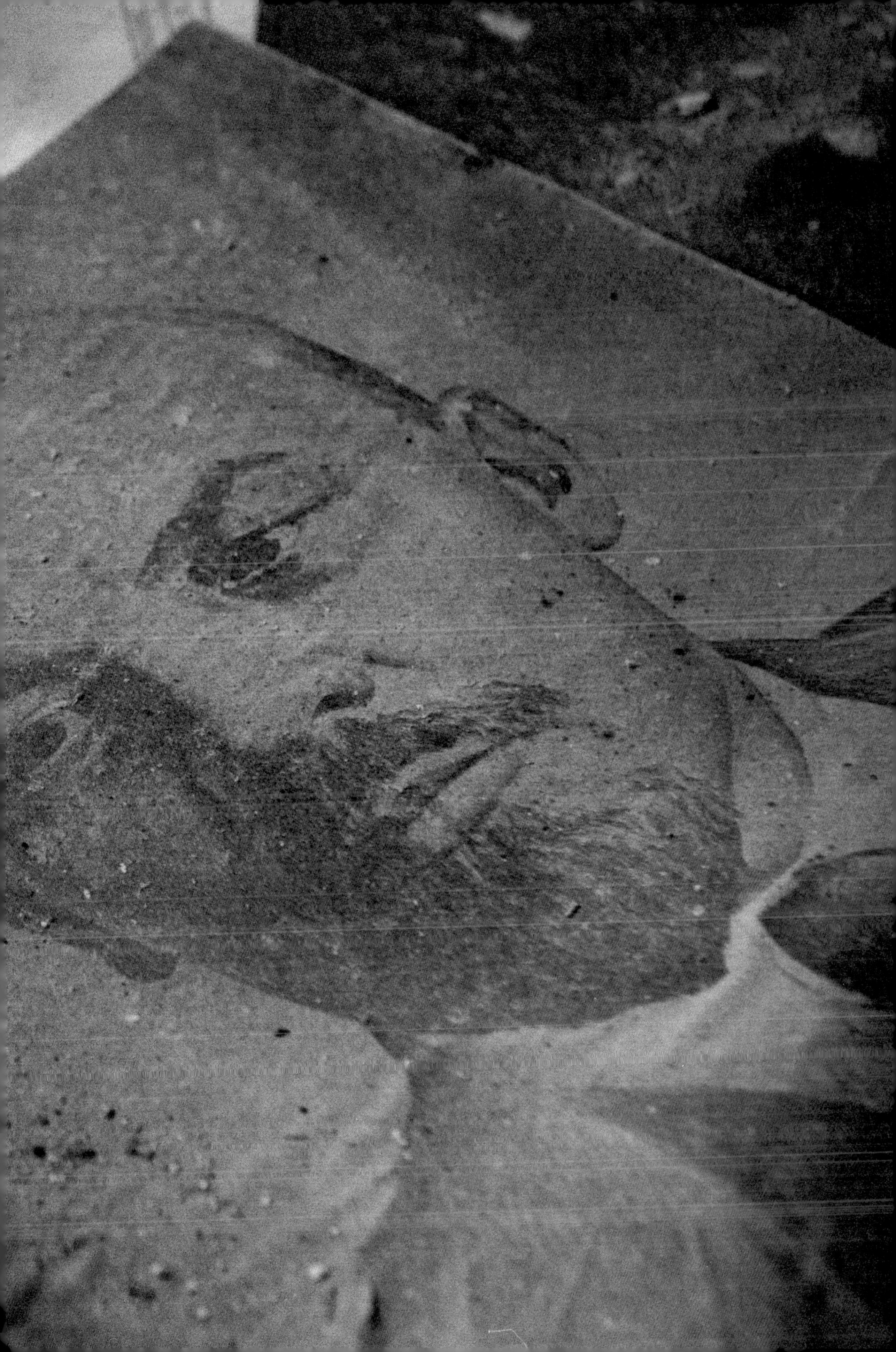

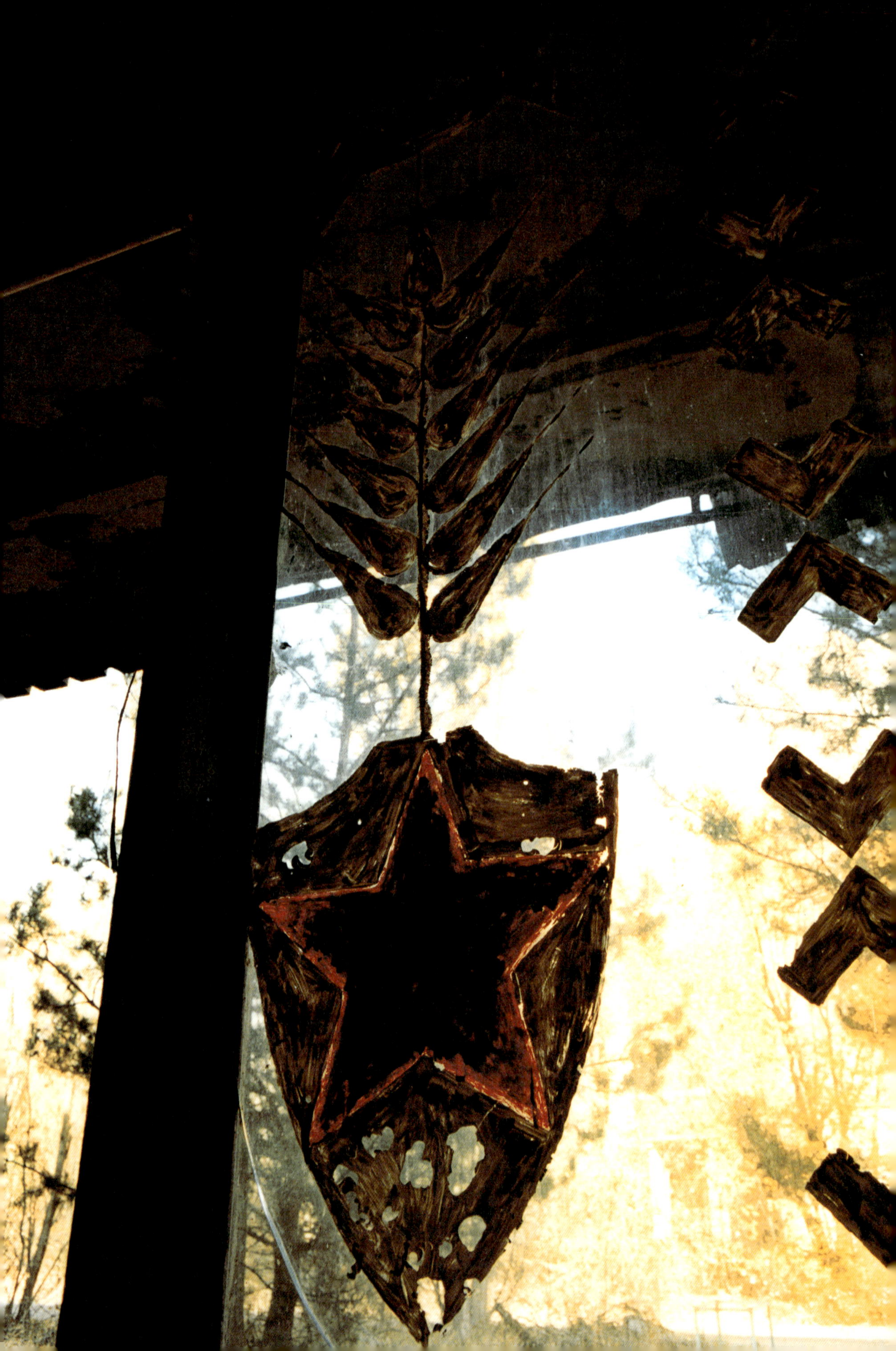

Confined the Soviet Union

Chernobyl power plant accident occurred just 1 year after the starting of the Gorbachev administration, and just after Perestroika was officially announced. According to the joint statement with U.S. president Reagan, ironically this accident triggered not only Glasnost but also the unity of West and East Germany, liberalization of the East European countries, and the corruption of the Soviet Union itself. Utopian country which Lenin had aimed to create departed from his ideal theory with the lapse of time. It is said that cause of nuclear power accident also stemmed from the corruption of the bureaucracy.

17 years have past from the collapse of the Soviet Union. But in Pripiati the state at the time of the accident remains even today. Anybody can see it clearly at a glance walking around the ruined town. Everywhere we can see country's symbol of sickle and hammer, national flag. In every public facility, slogans which hail nationalism and every day's labor were painted on the walls. We can find Lenin everywhere on the wall and on the pages of many books. He is as if protecting the ruined town.

A man rose up to create a Utopian country where people can lead a humble life equally. His policy which hailed national electrification throughout the country for the people's happiness led to the motivation to bear advanced nuclear technology. But how does he conclude this reality as a result that very technology itself ruined the town?

封印されたソ連

　チェルノブイリ原発事故はゴルバチョフ政権が発足してわずか１年後に発生した。レーガン米大統領との共同声明に基づき、ペレストロイカ（改革）を提唱した矢先の出来事であった。皮肉なことにこの事故をきっかけにグラスノスチ（情報公開）路線も提唱され、東西ドイツ統一、東欧衛星国家の自由化、果てはソ連邦自身の解体の遠因となったともいわれている。ウラジーミル・レーニンが目指したユートピア国家は時の経過とともに、初期の理想から大きく外れ様々な歪みの病巣を抱え込んでいた。発電所事故の根本の原因も腐敗した官僚主義による無理な実験にあったといわれている。

　ソ連崩壊から17年。しかしプリピアチの町は今もなお事故当時のソ連がフリーズされている。廃墟の町を歩けばそれは一目瞭然だ。至る所に配された国家のシンボルマークである鎌、トンカチ、工で星。公共施設には国家高揚、労働讃歌のスローガンが壁画で描かれ、学校には戦意高揚のポスターが貼り付けられたままだ。そして至る所にウラジーミル・レーニンがいる。あるときは肖像画で、あるときは壁画で、またあるときは書物の１項となって……。あたかもこの廃墟の町を守っているようである。

　腐敗した帝政を打破するがため立ち上がった１人の男は、日々ささやかに生きる人間が平等に過ごせるユートピアを作ろうとした。幸福な生活のための国家電化をスローガンにした彼の政策は、後に高度な原子力技術を生む原動力となった。しかし結果、その技術が仇となってひとつの町が消滅したという皮肉な事実を、彼はどのように見ているのだろうか？

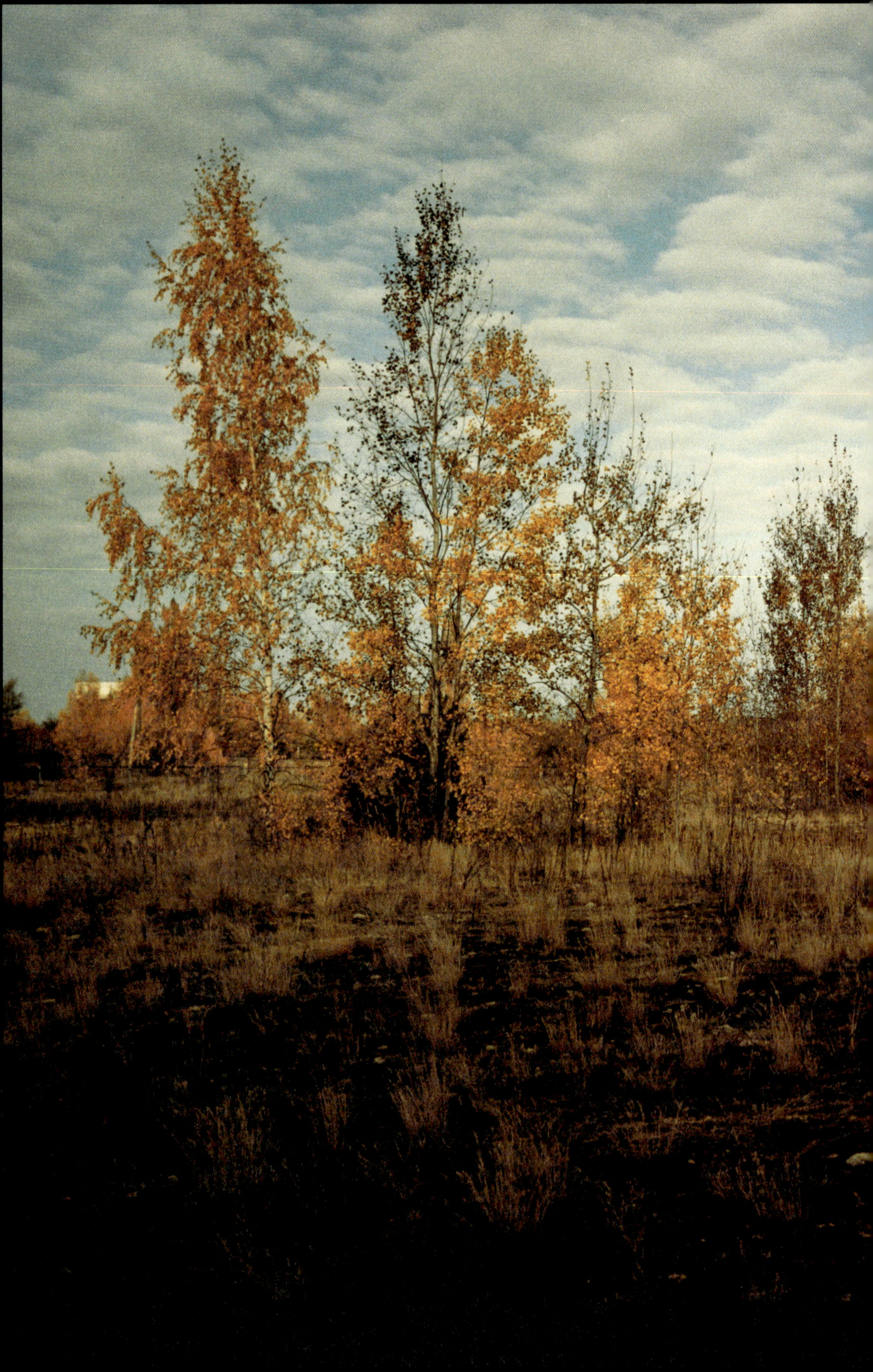

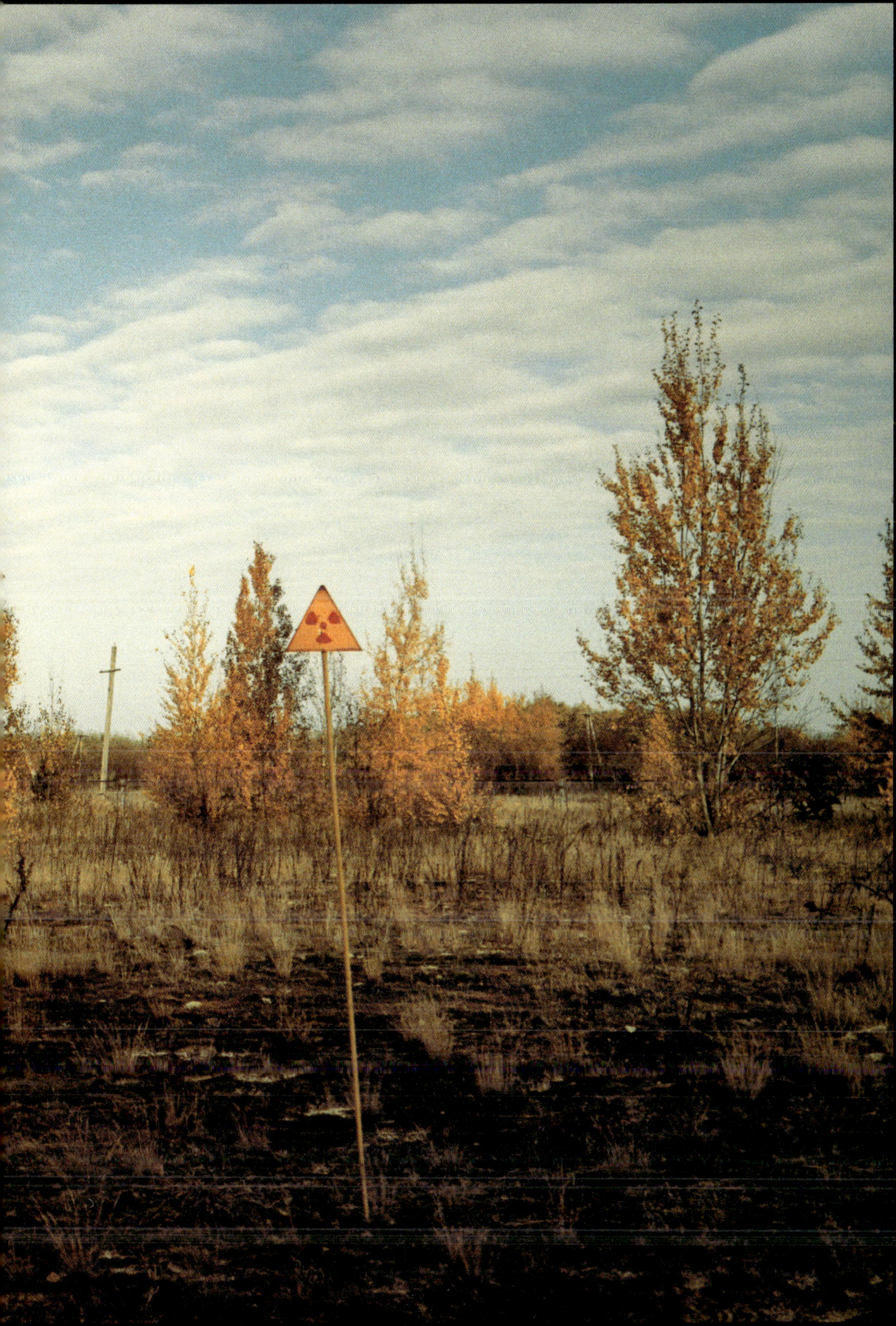

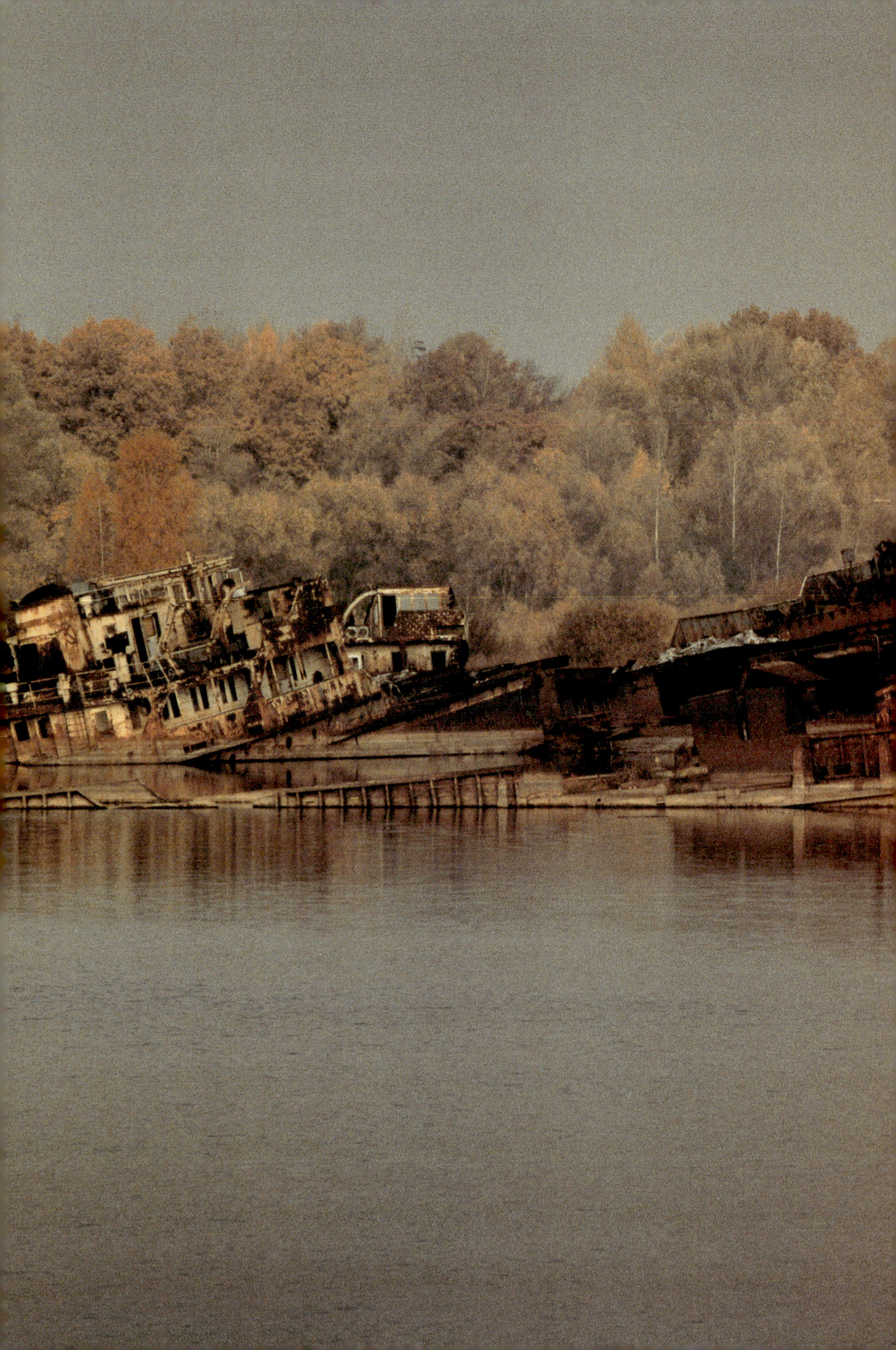

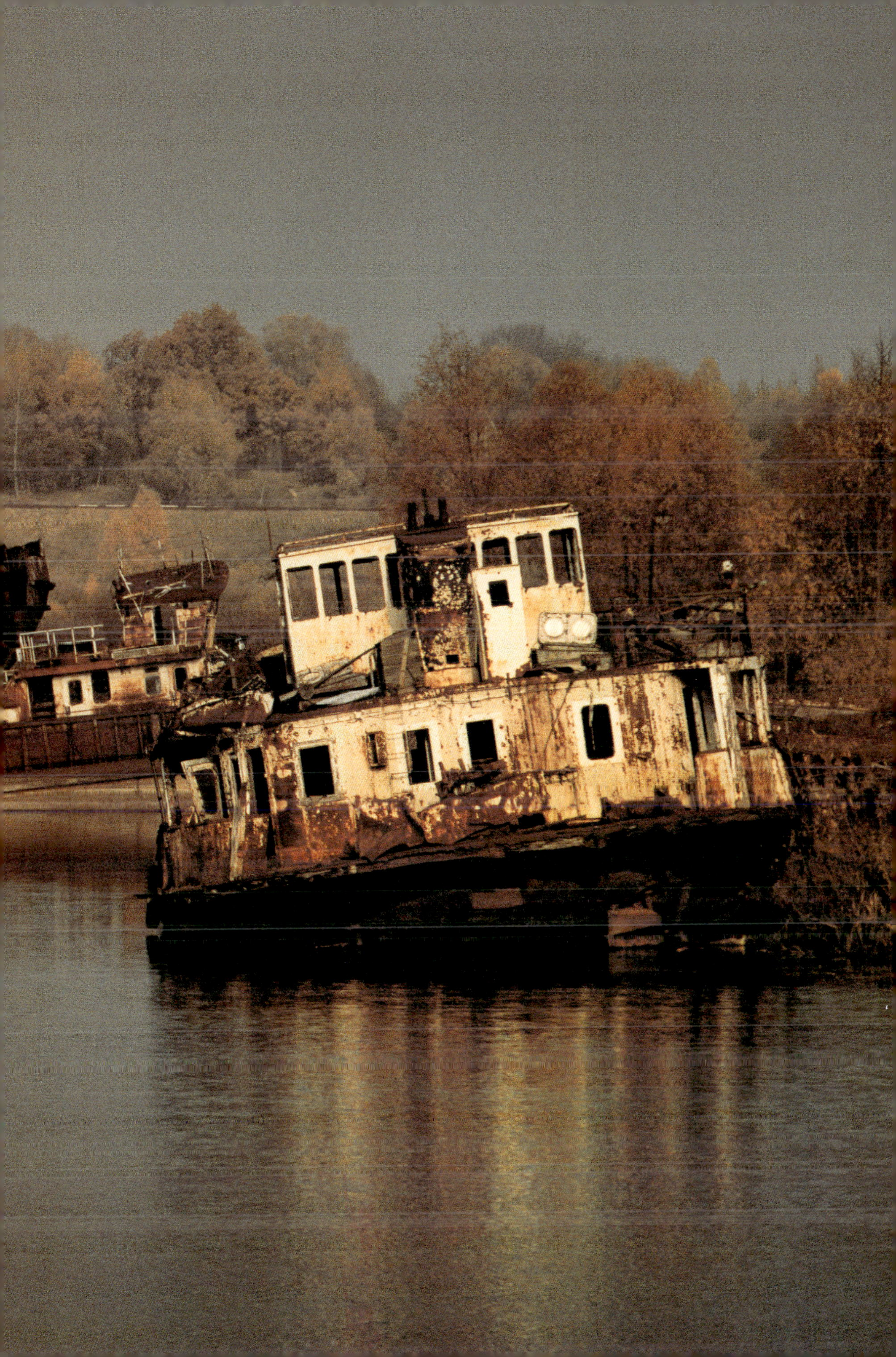

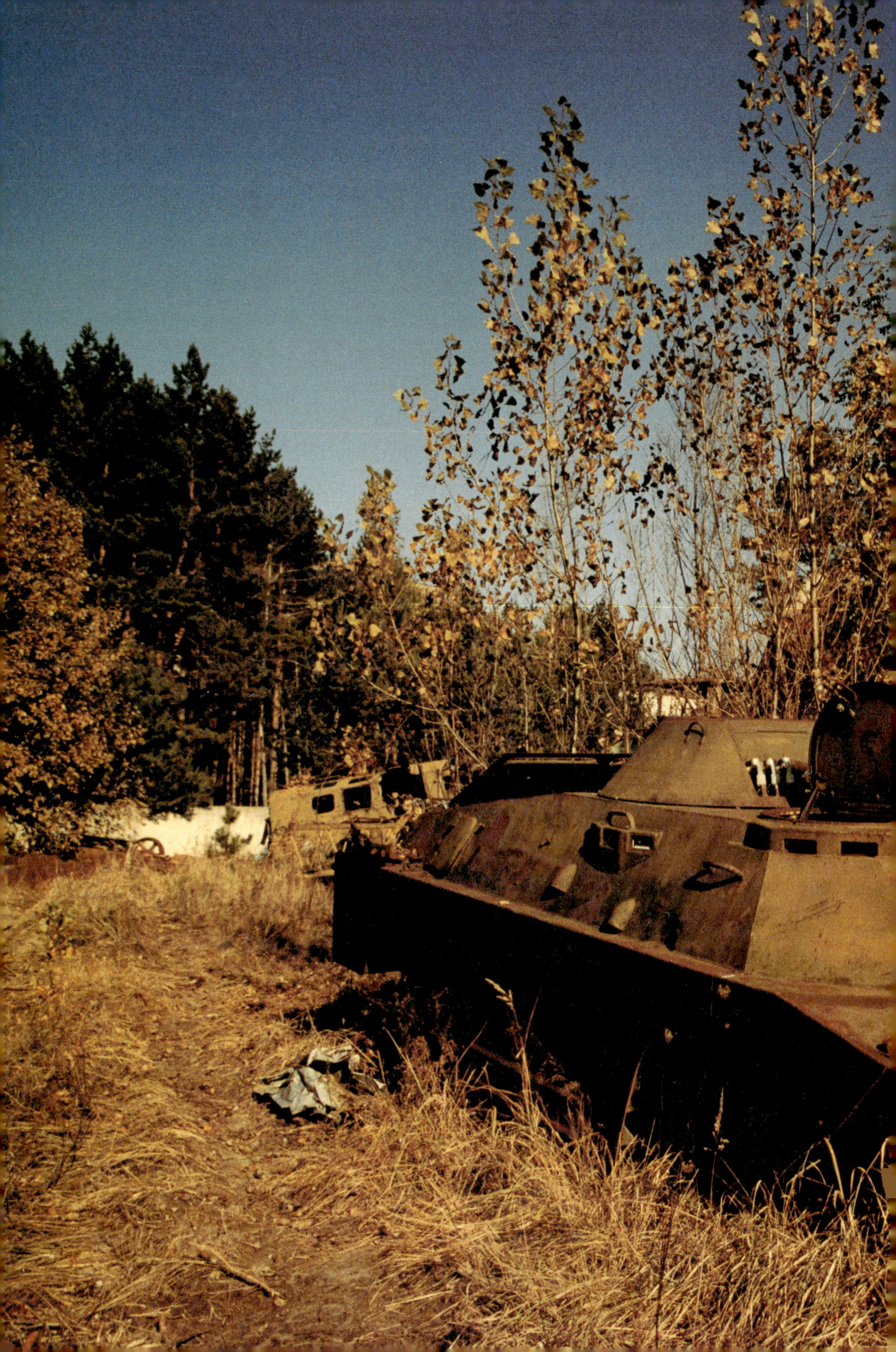

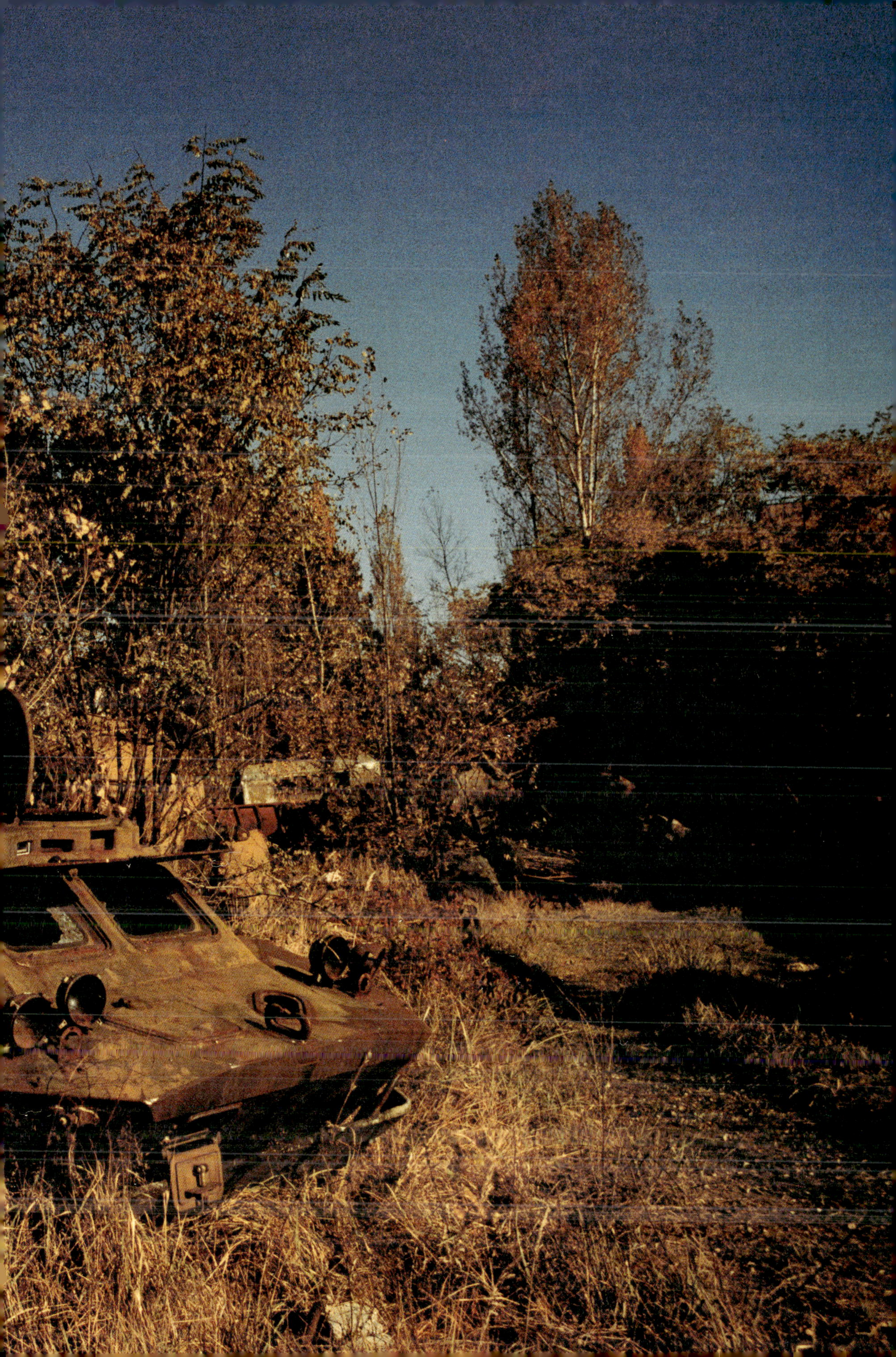

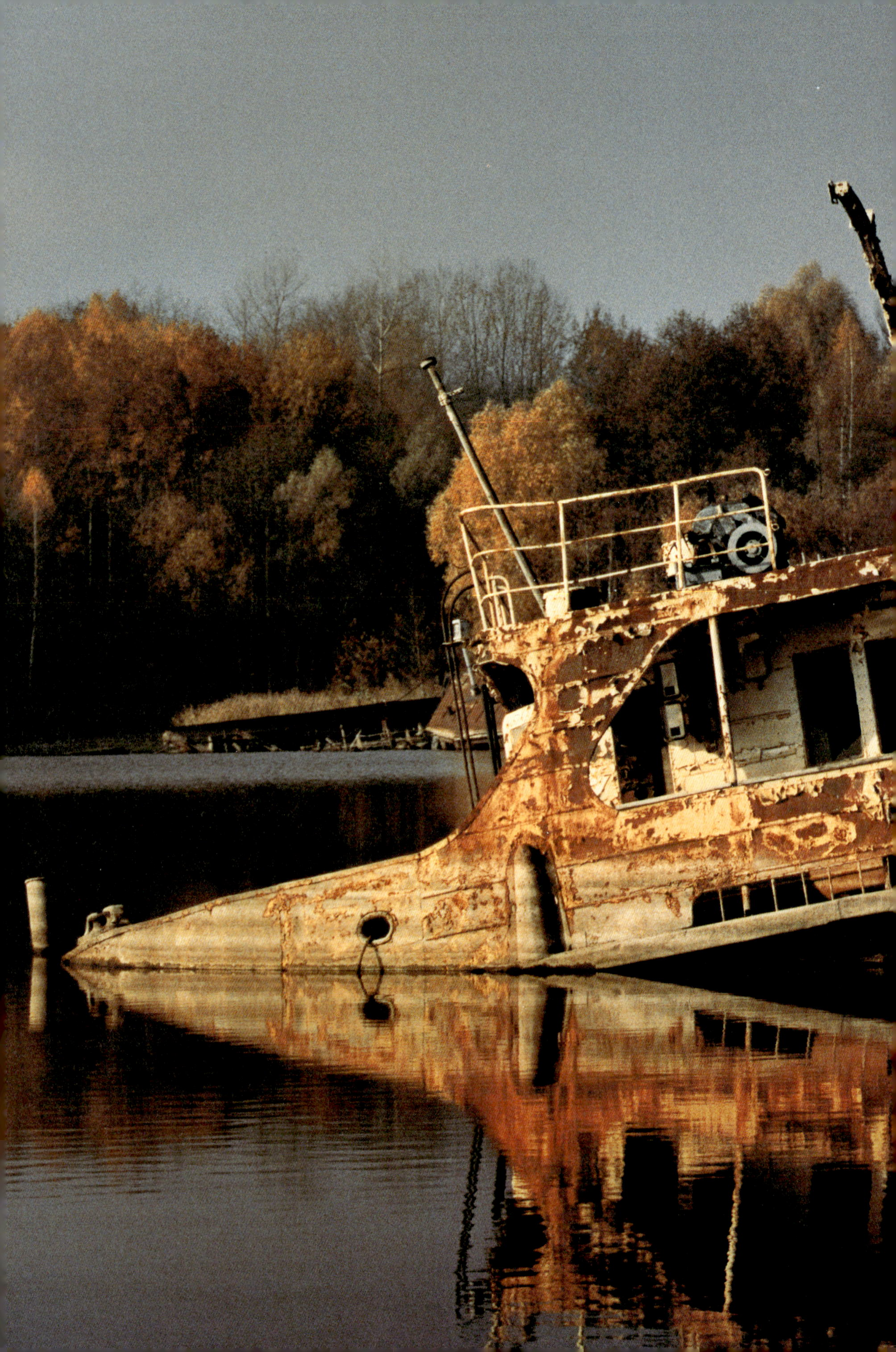

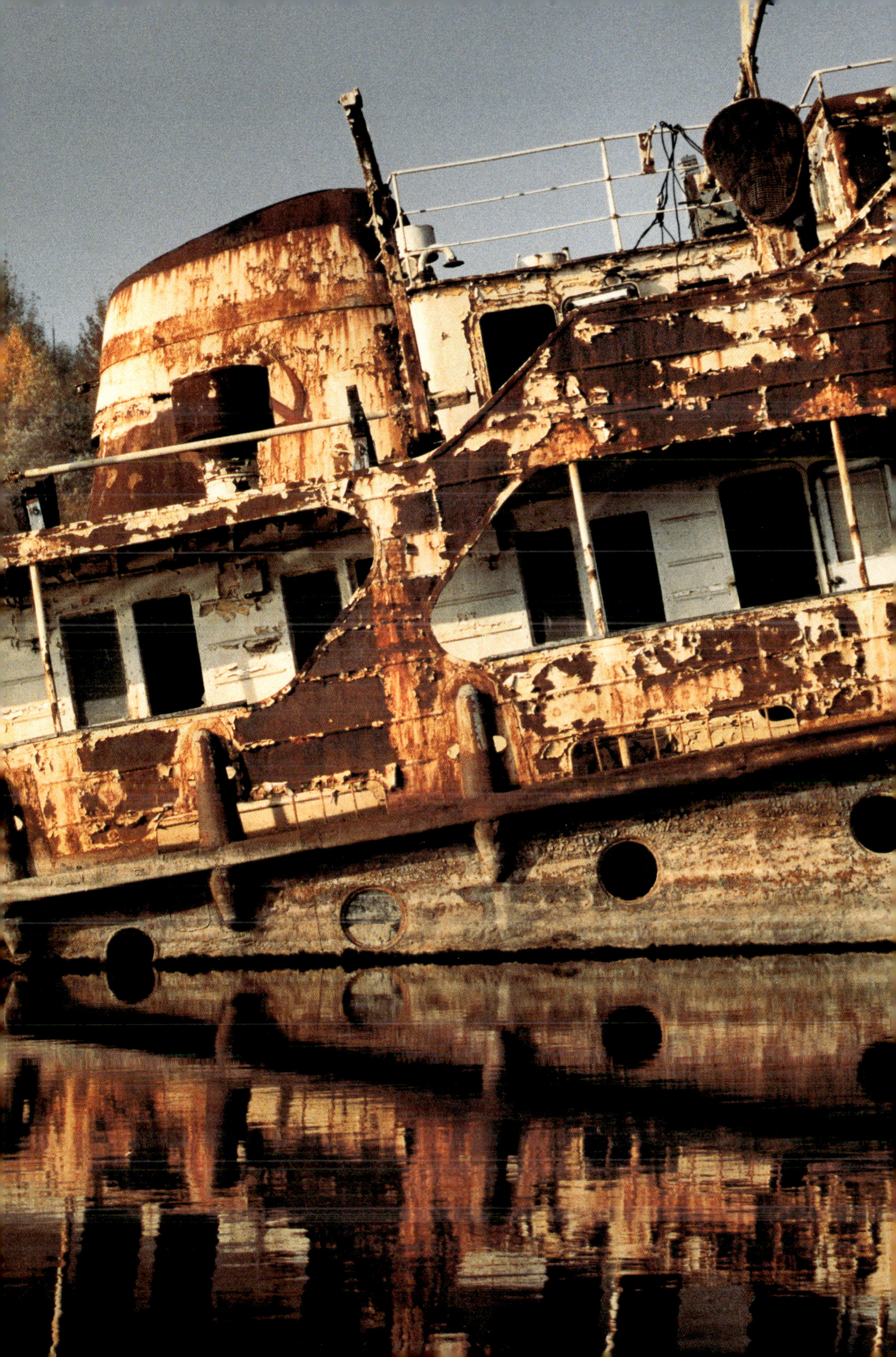

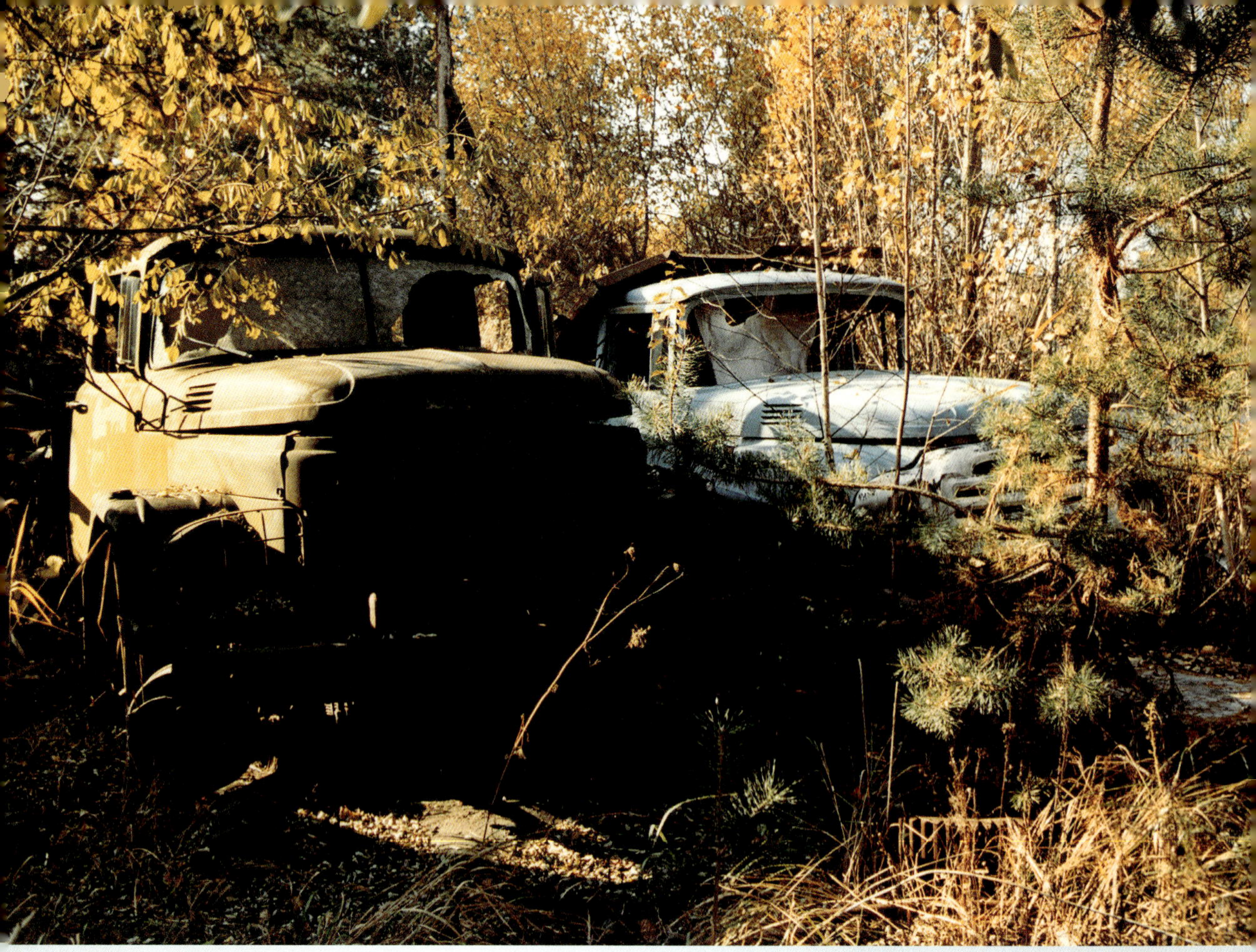

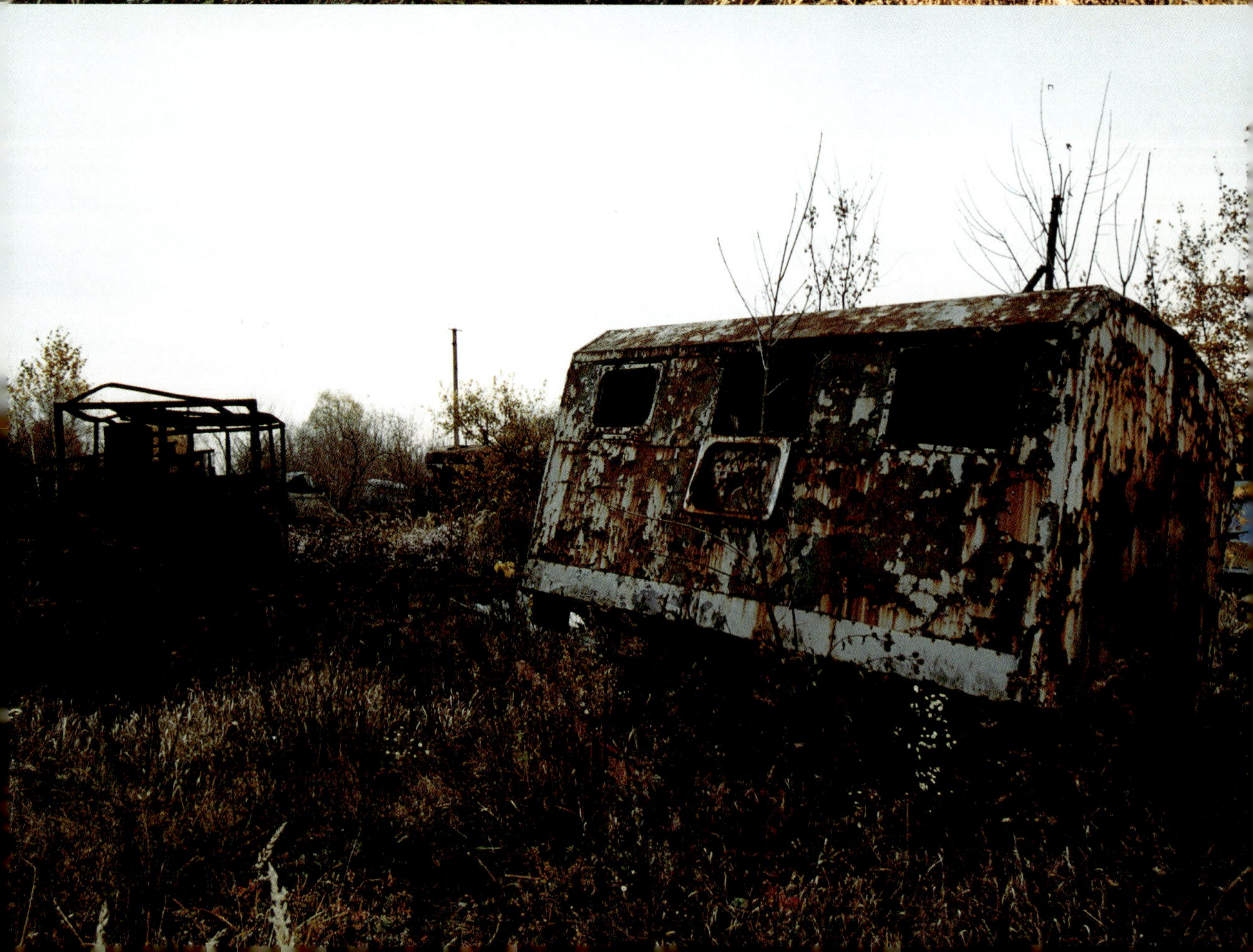

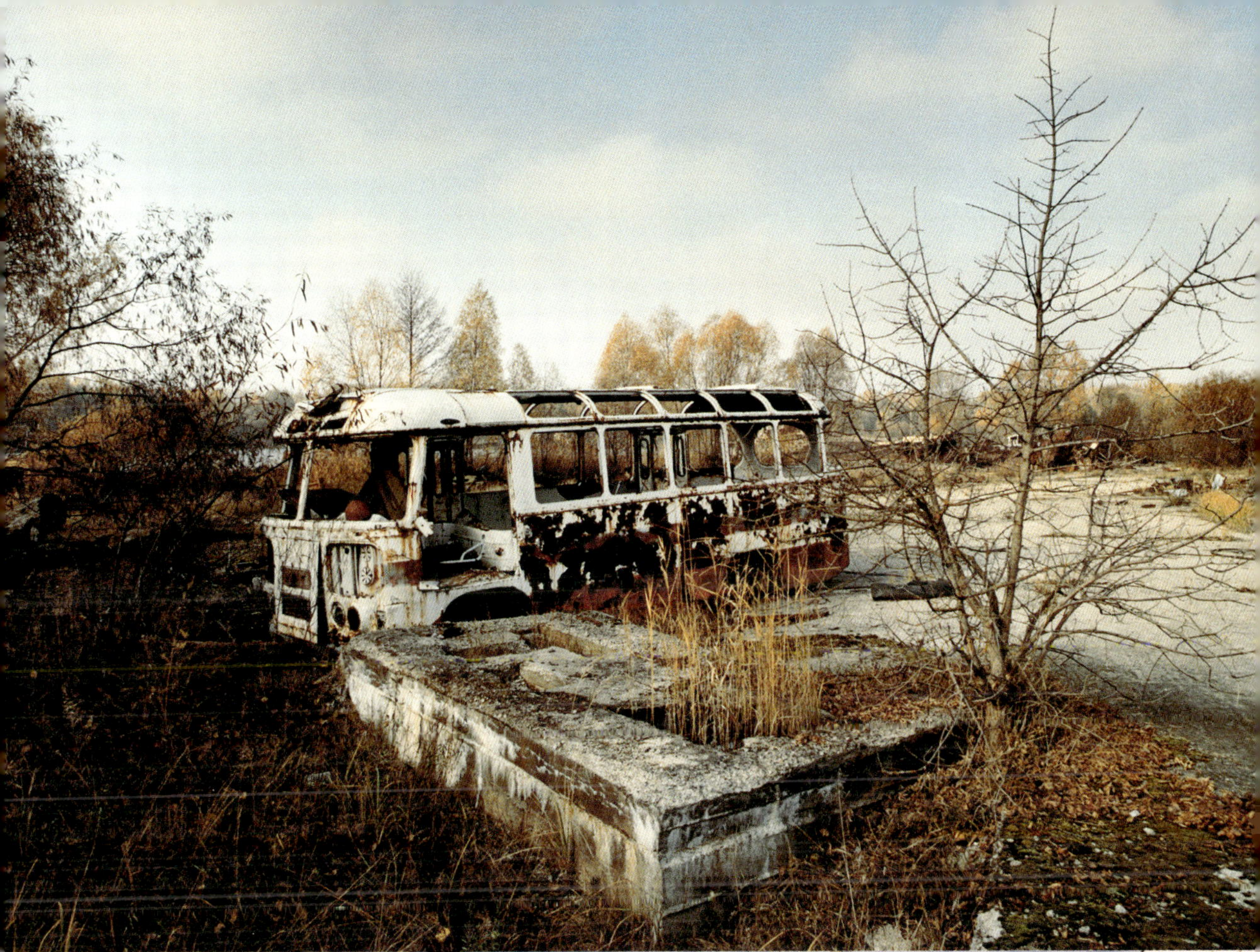

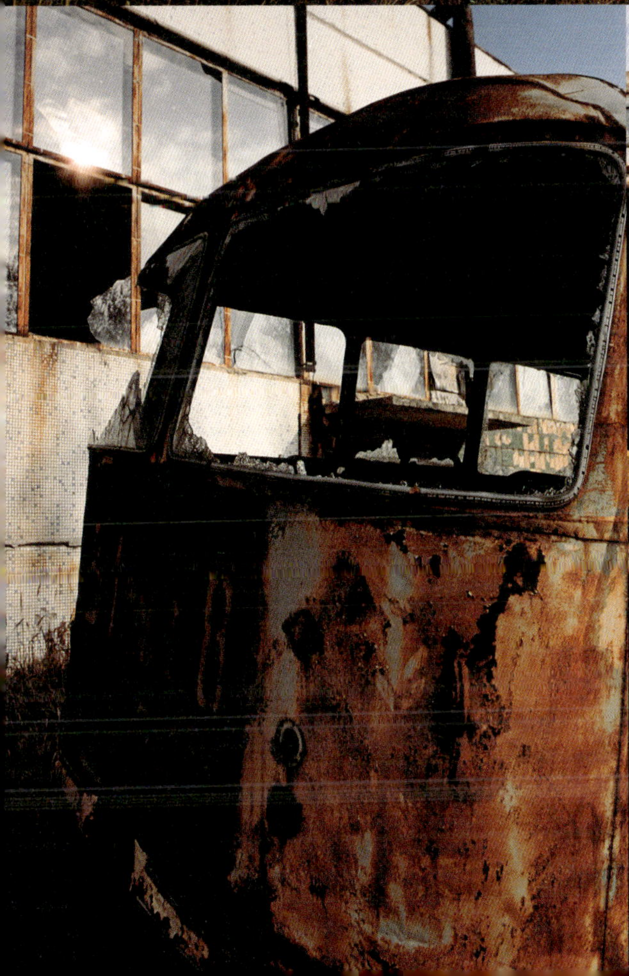

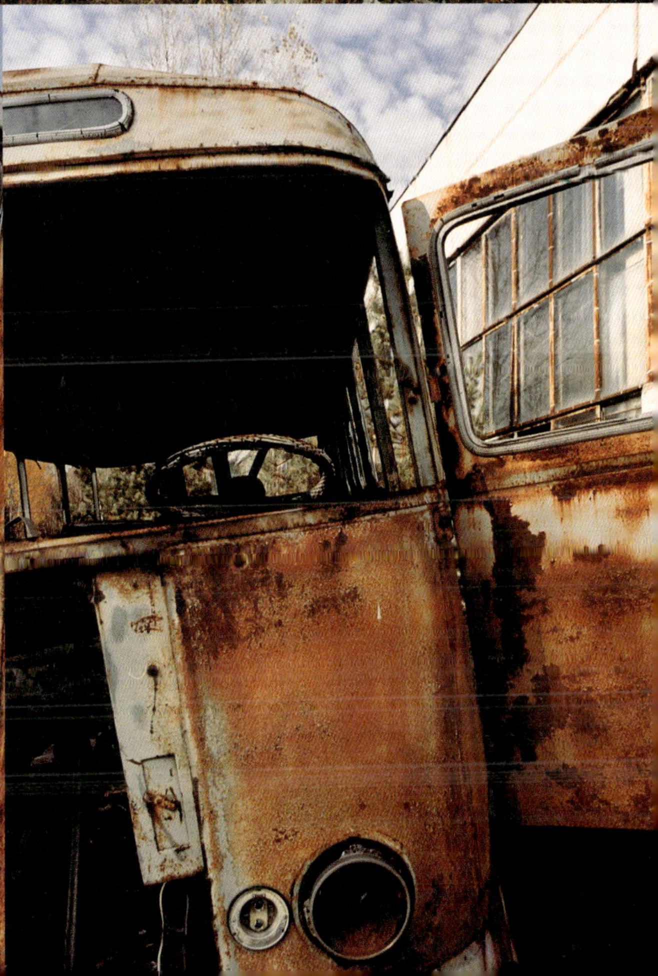

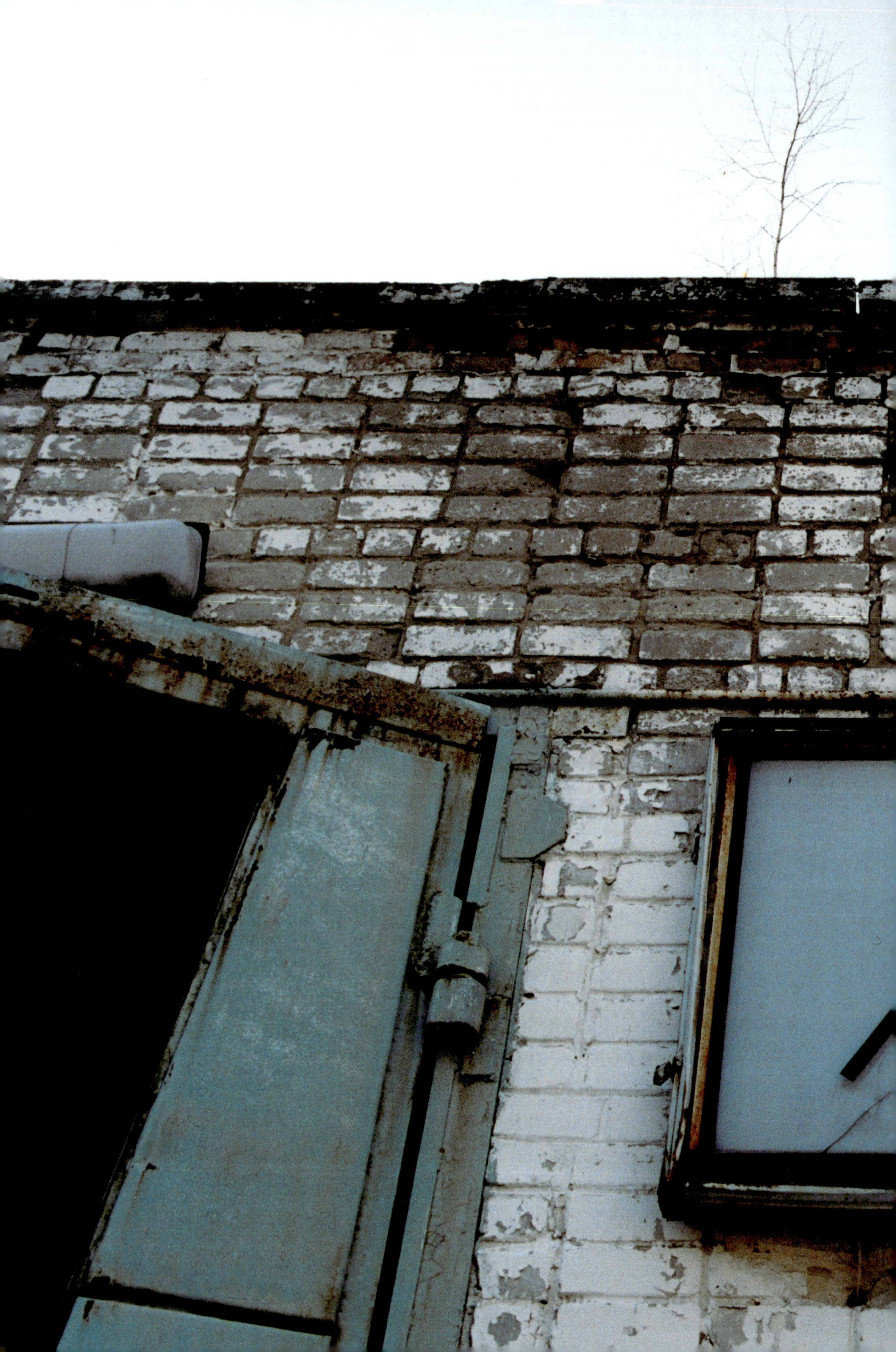

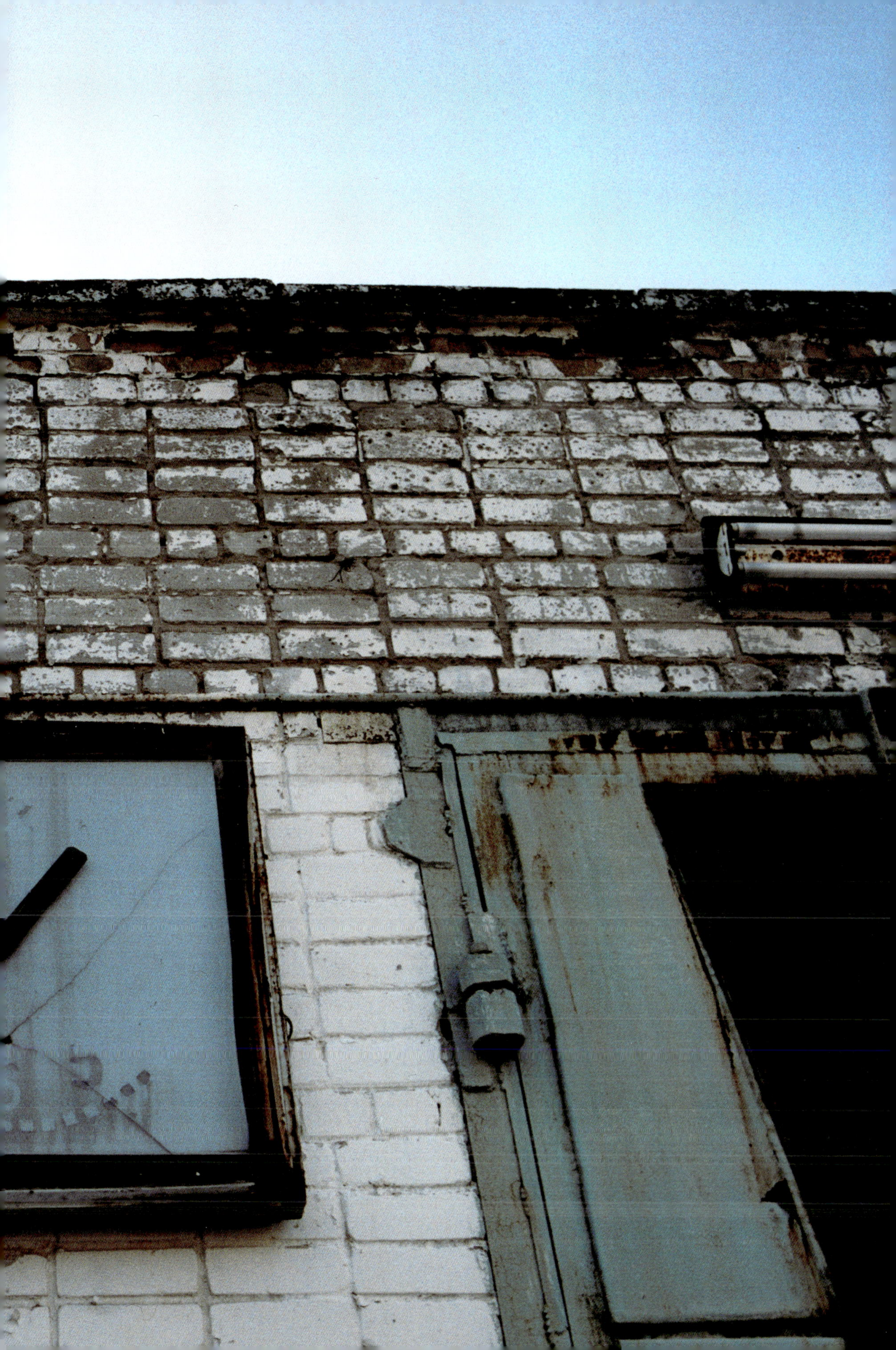

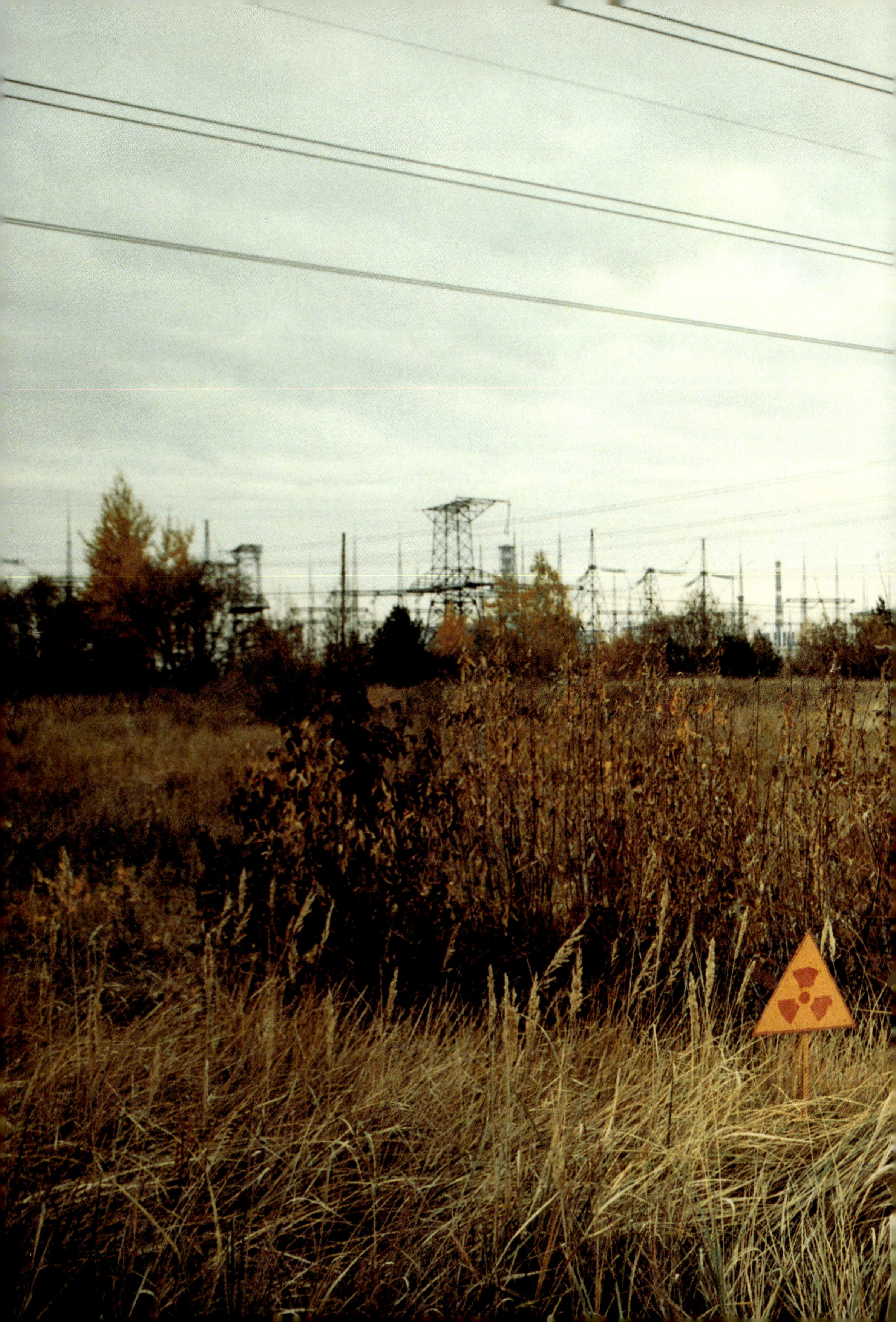

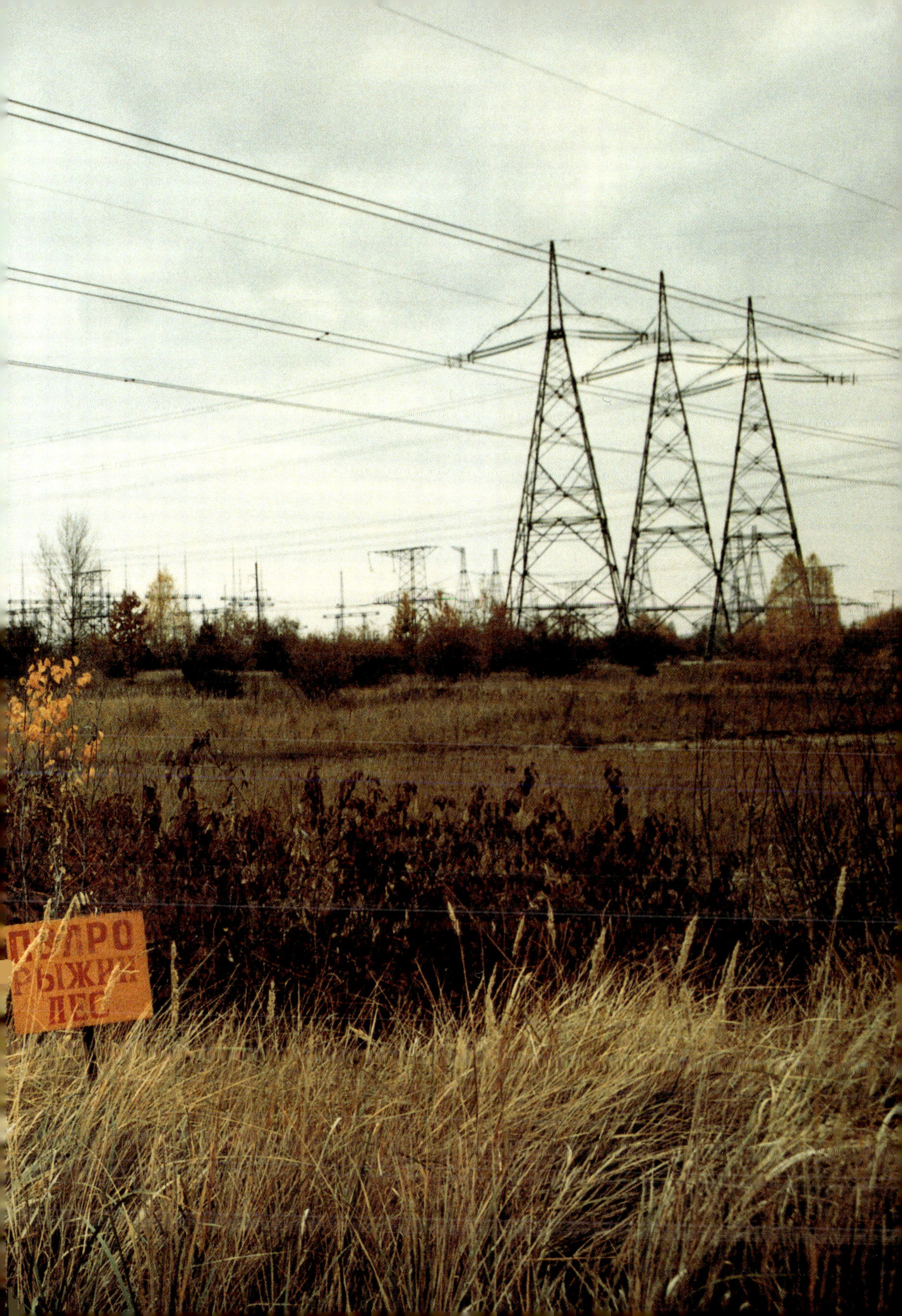

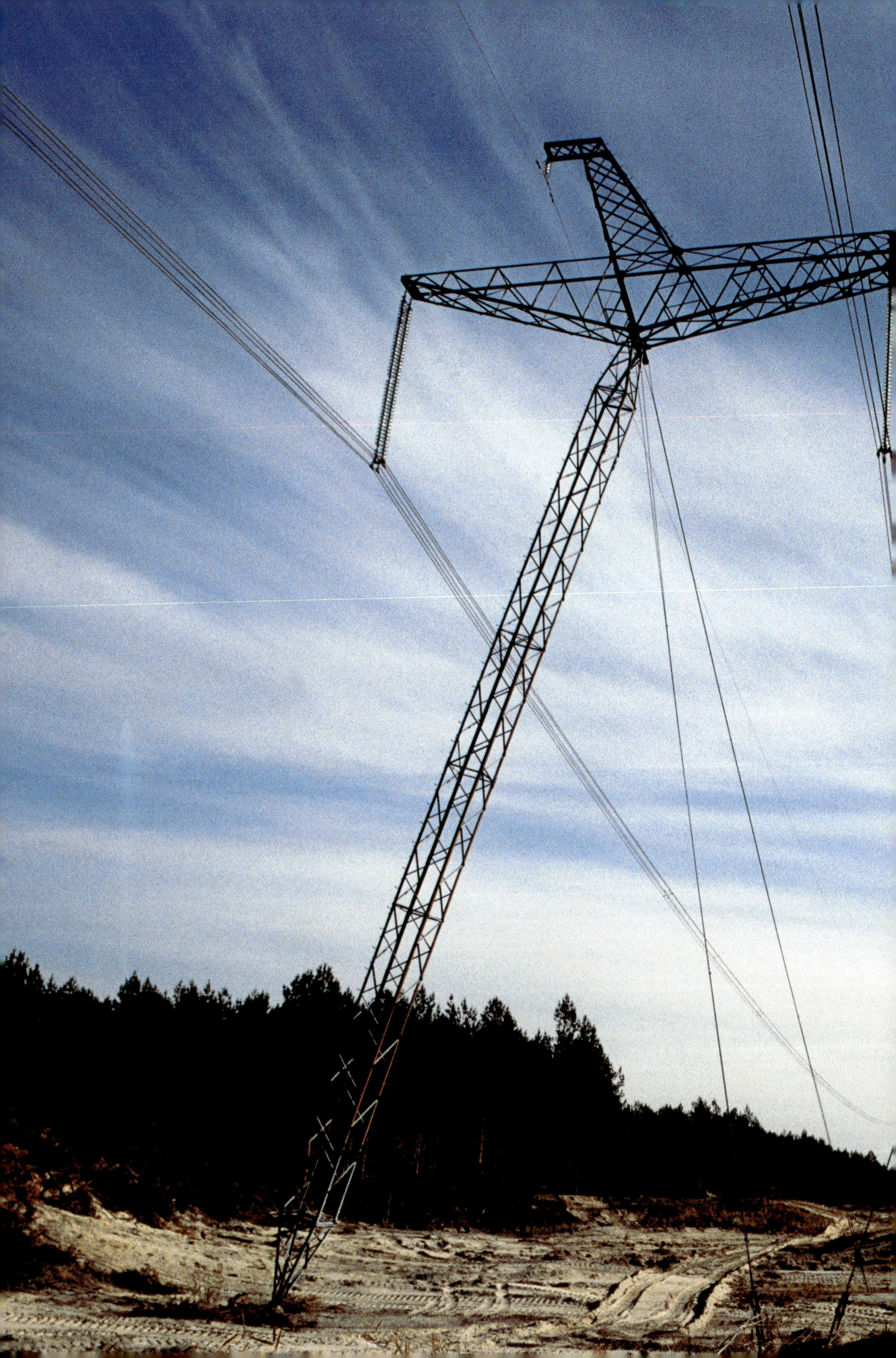

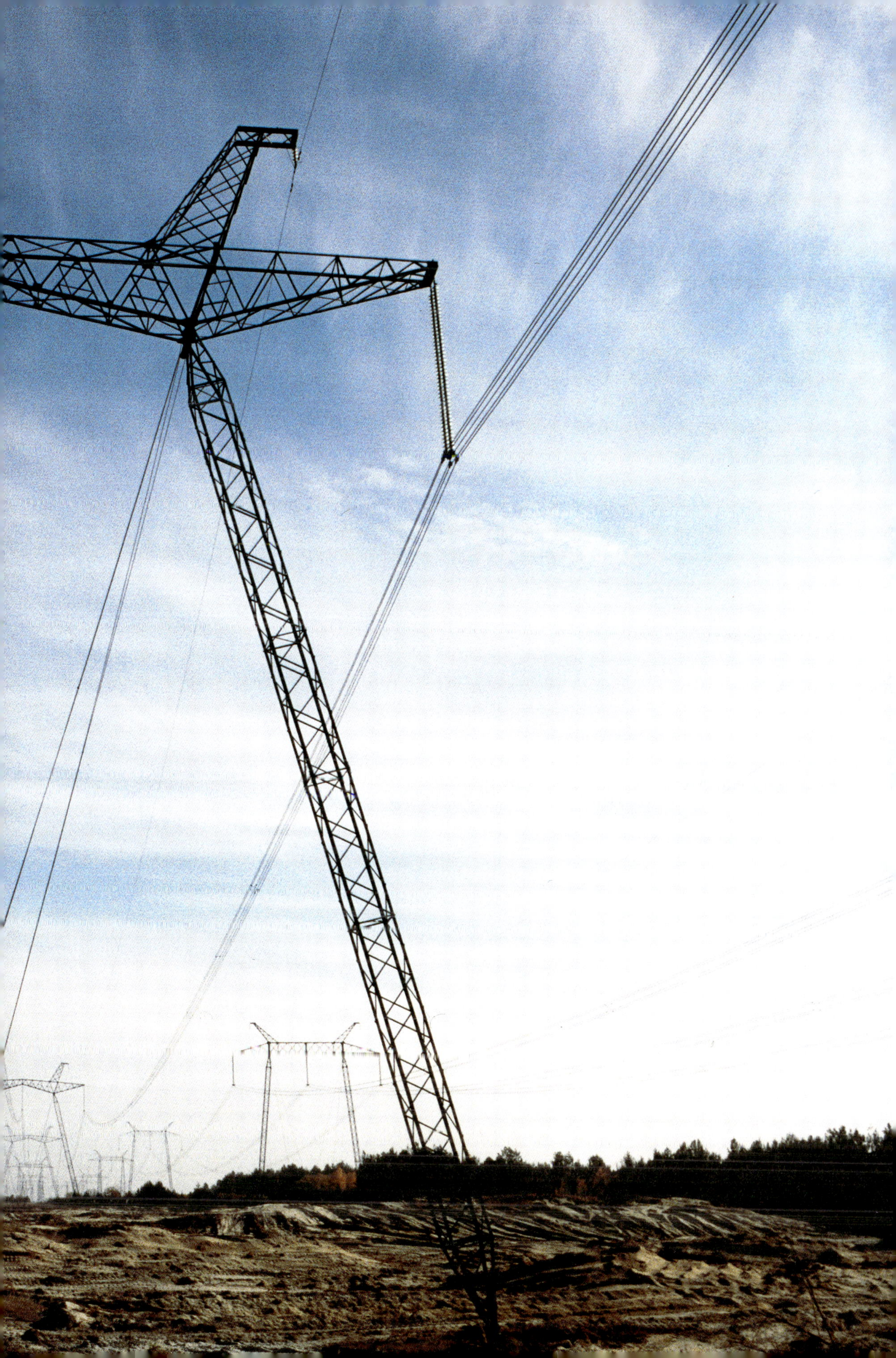

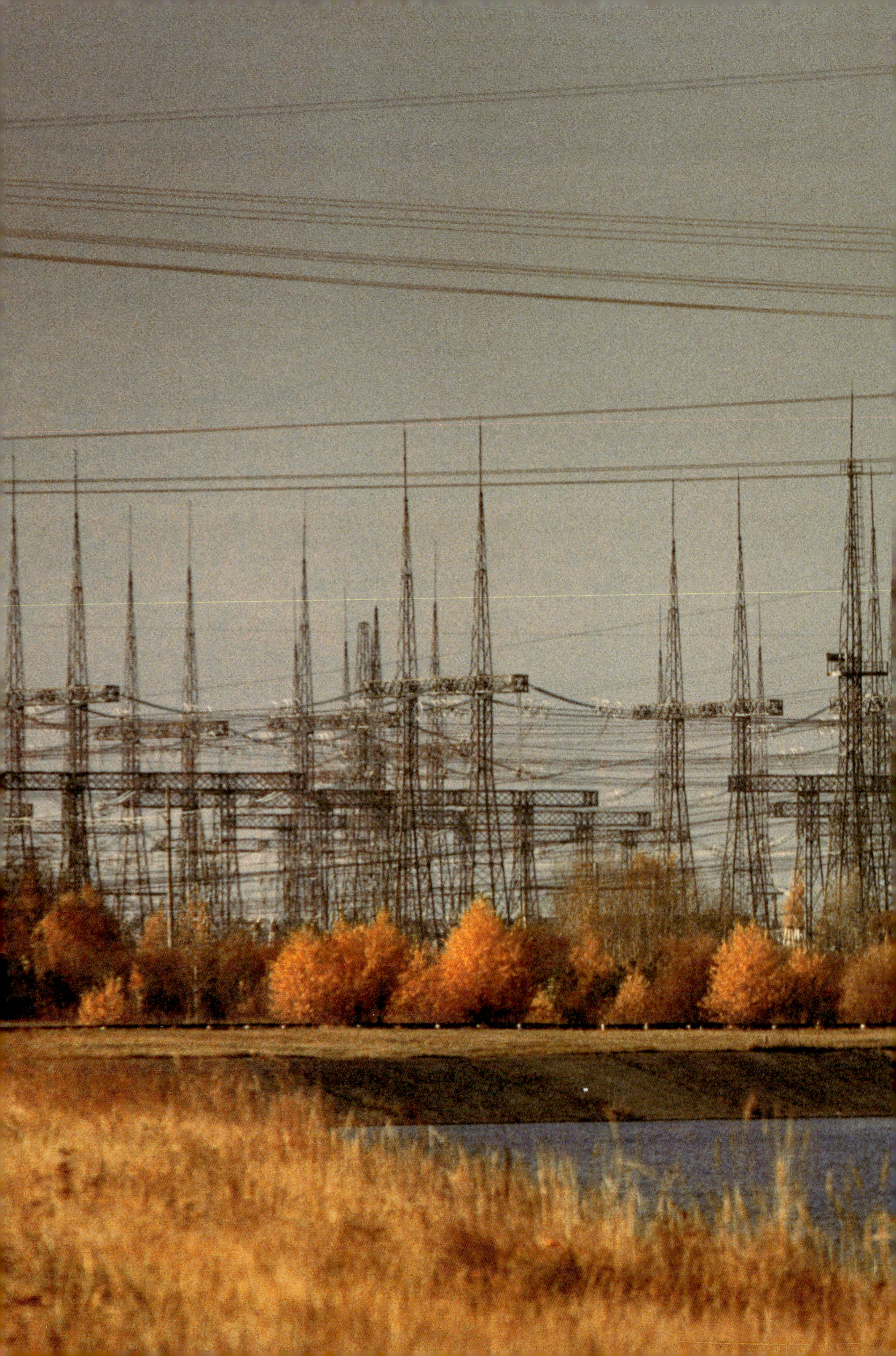

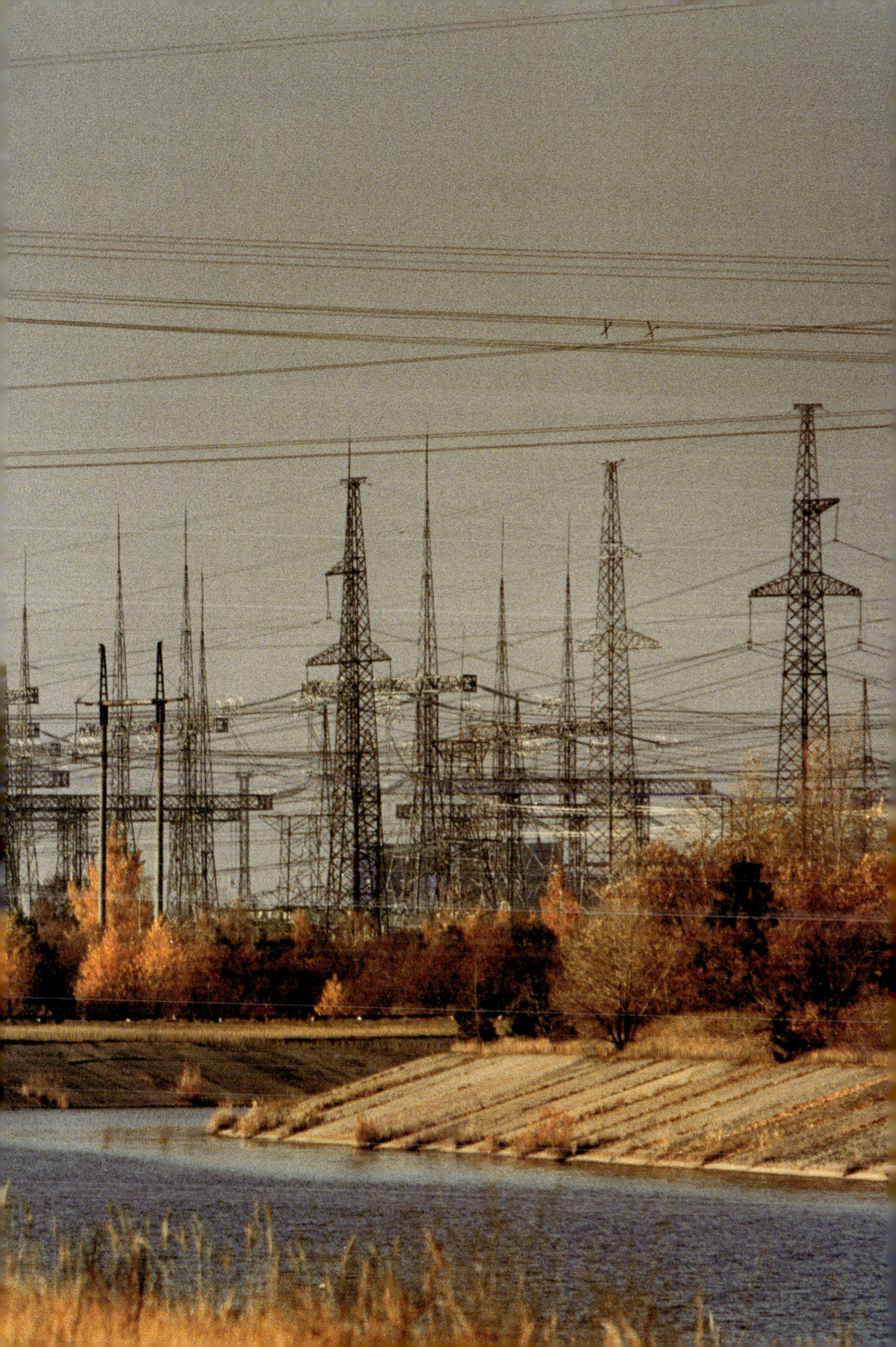

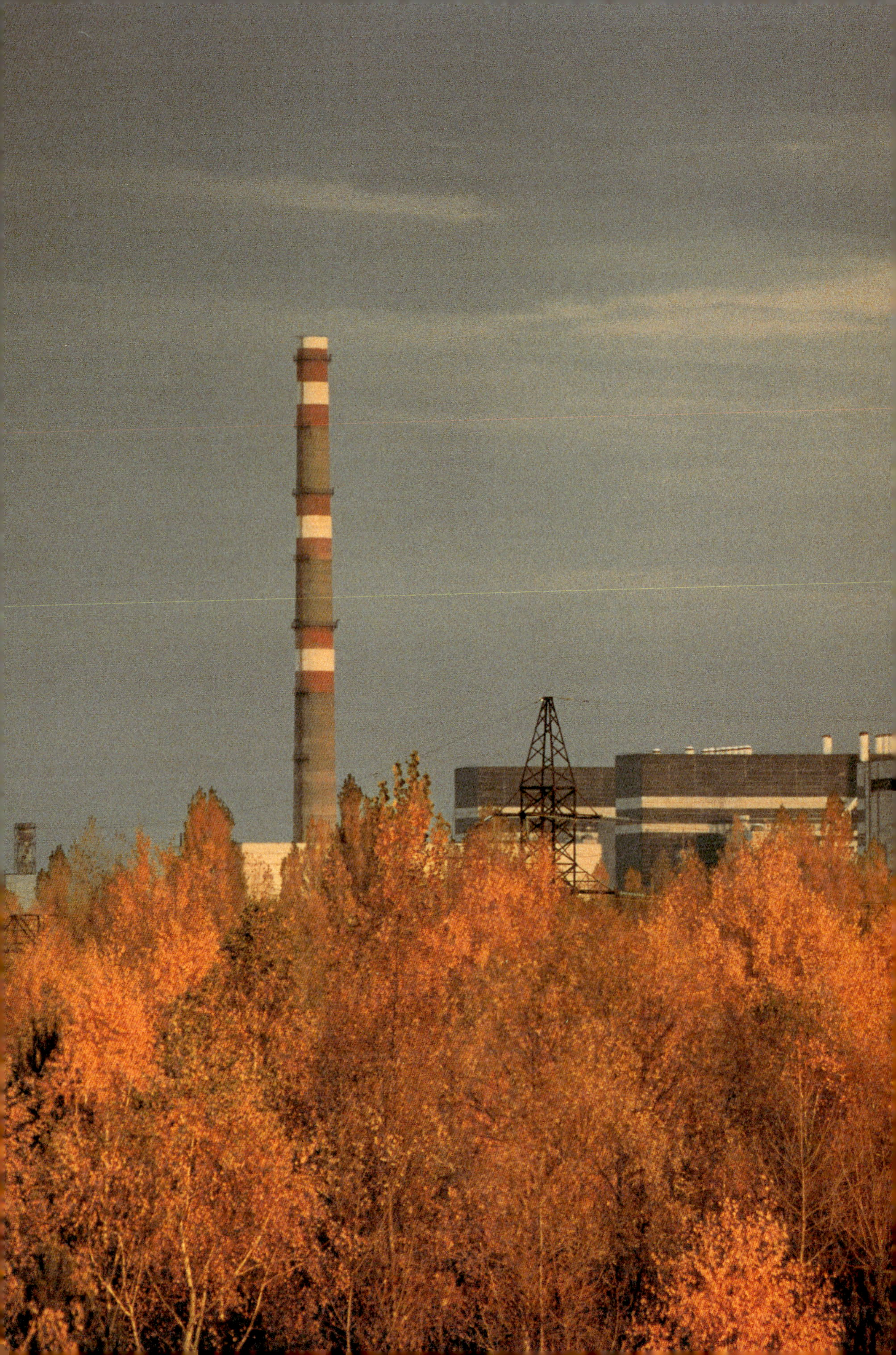

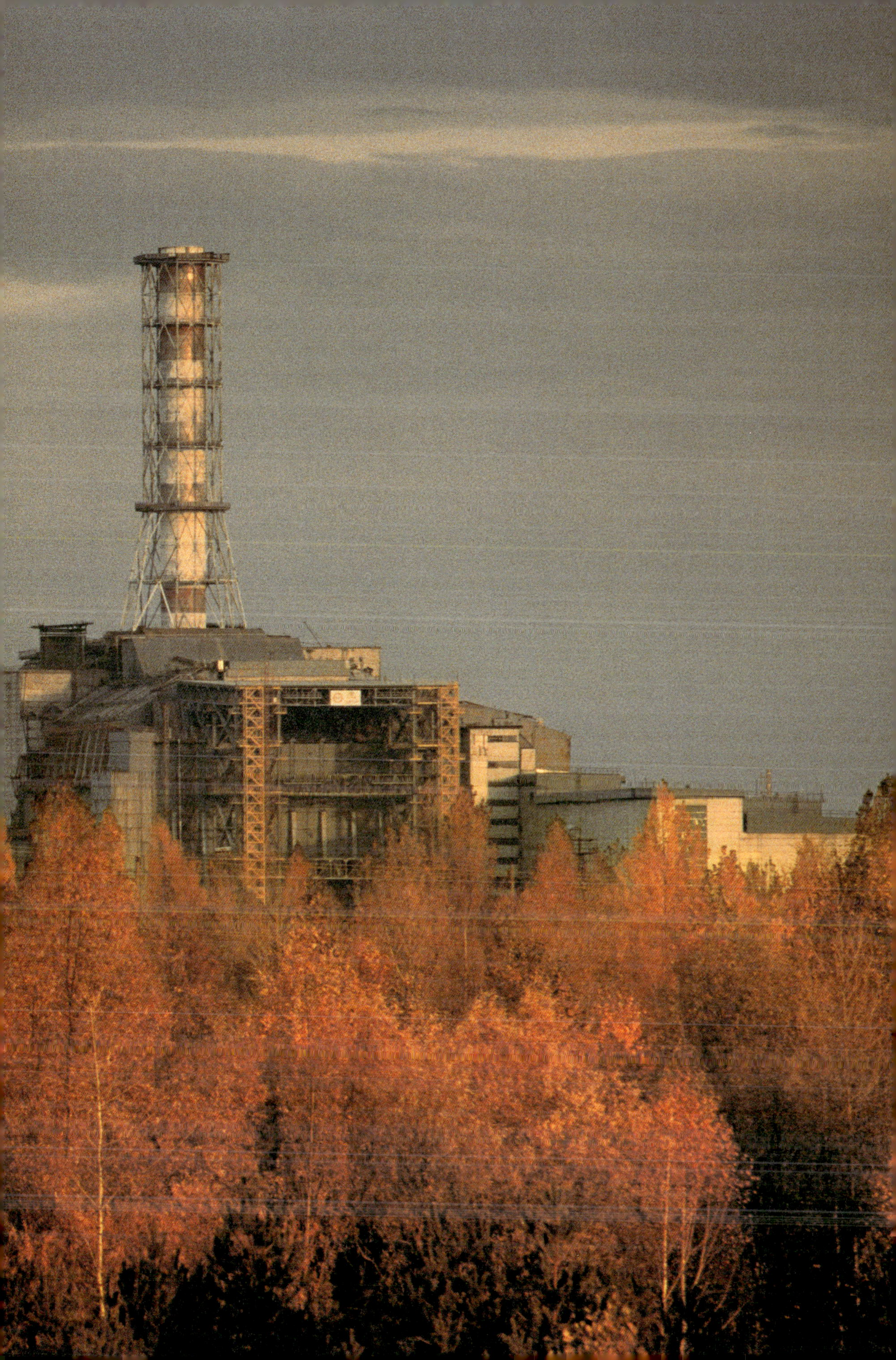

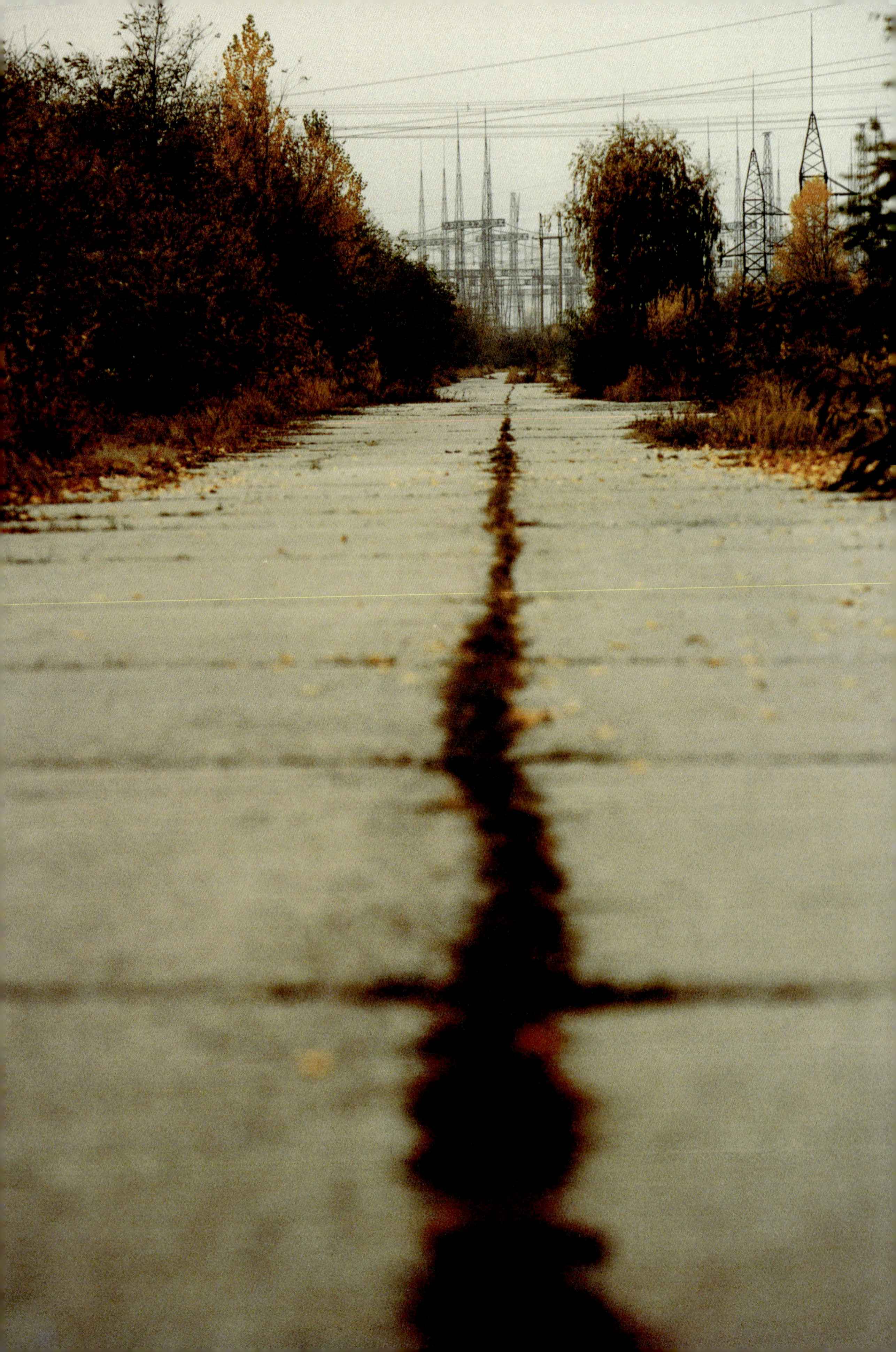

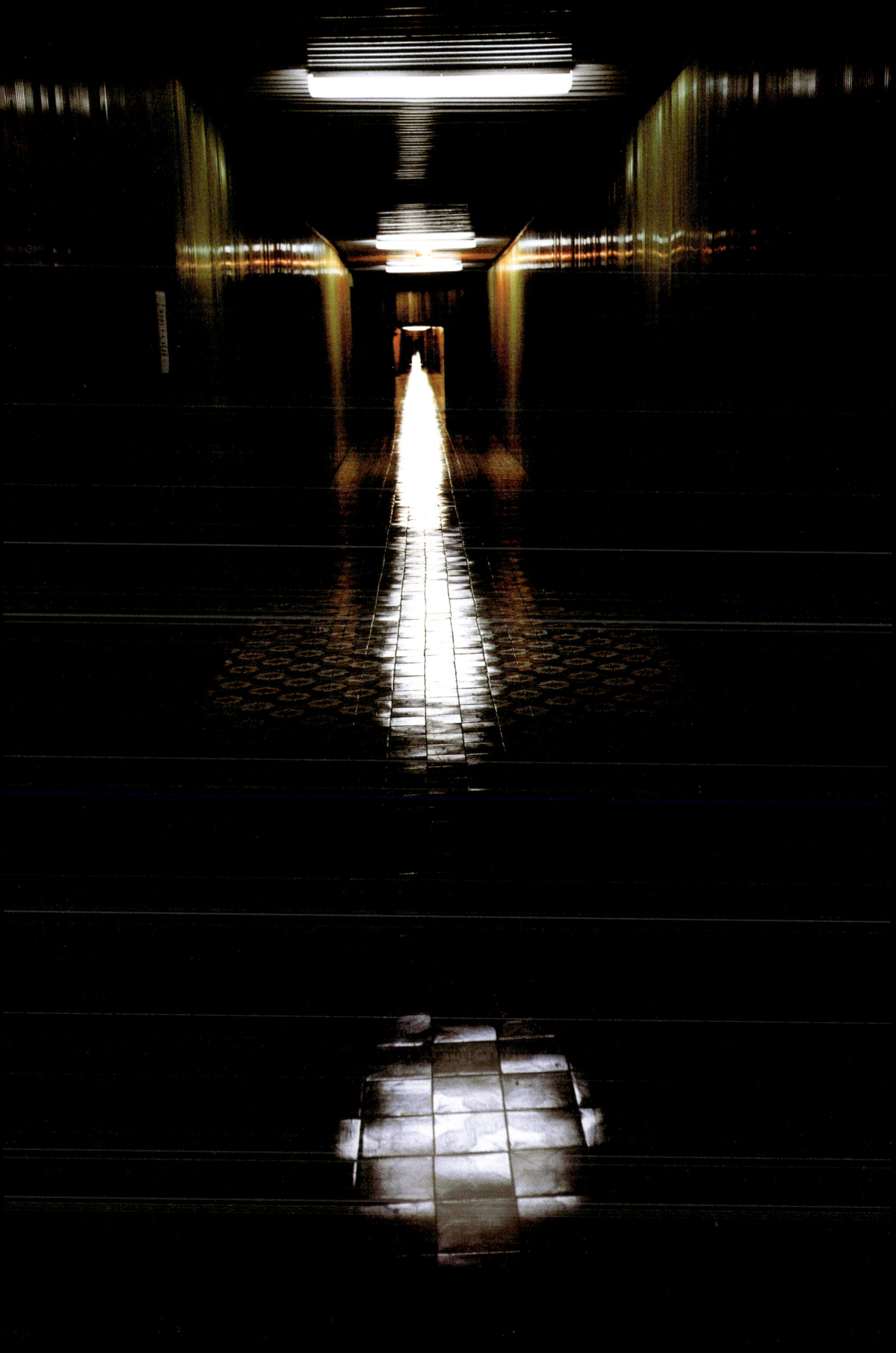

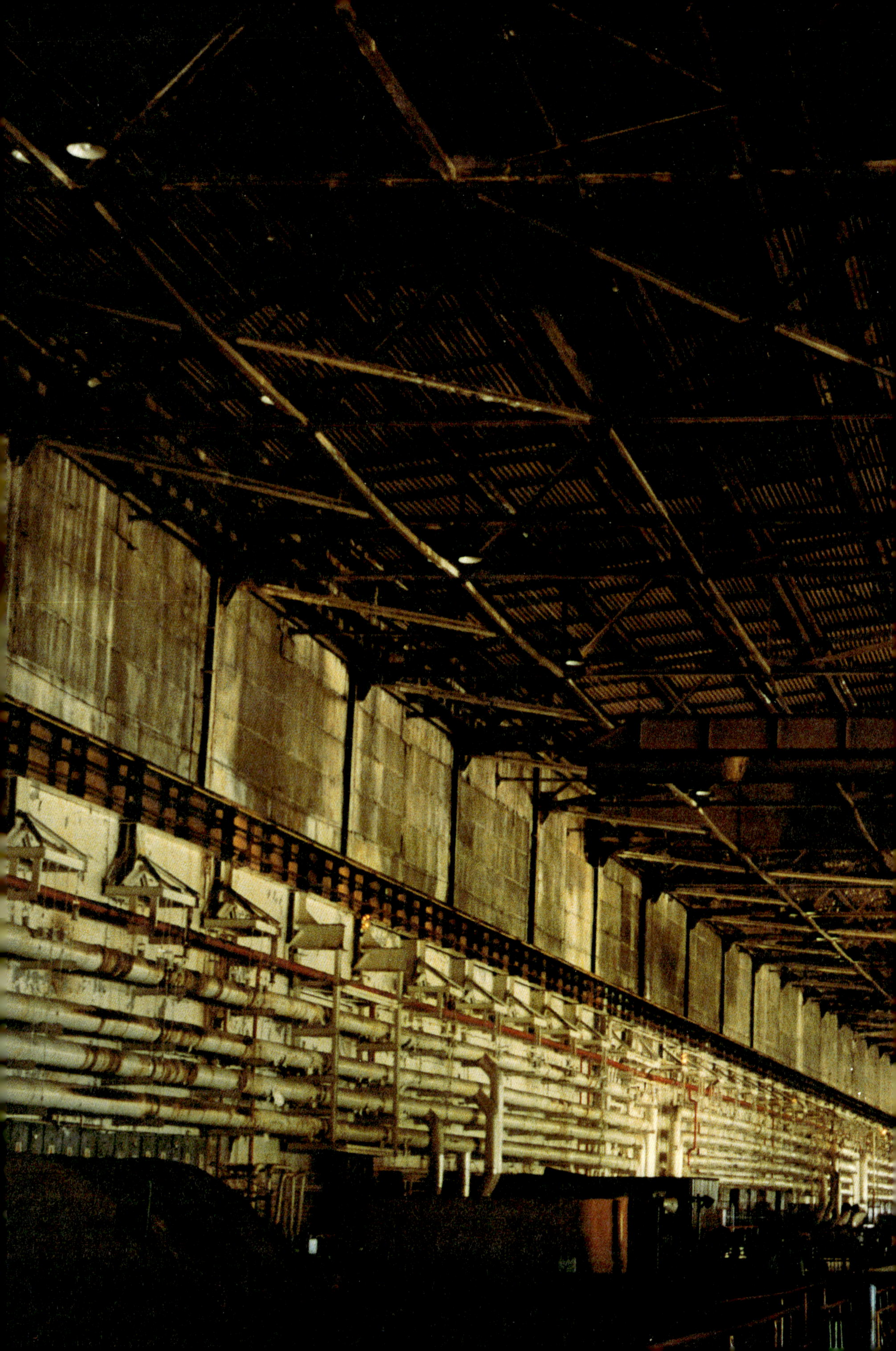

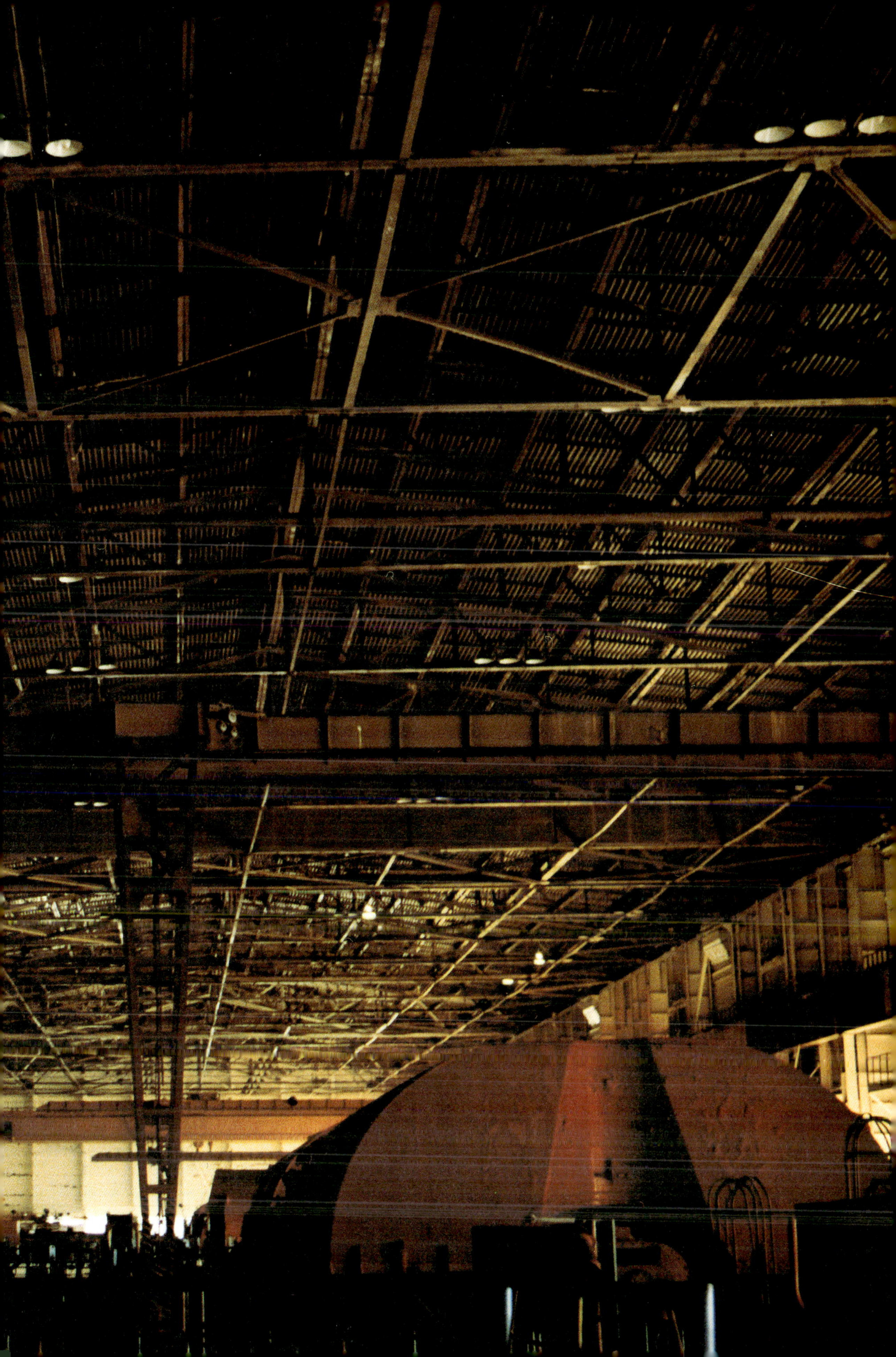

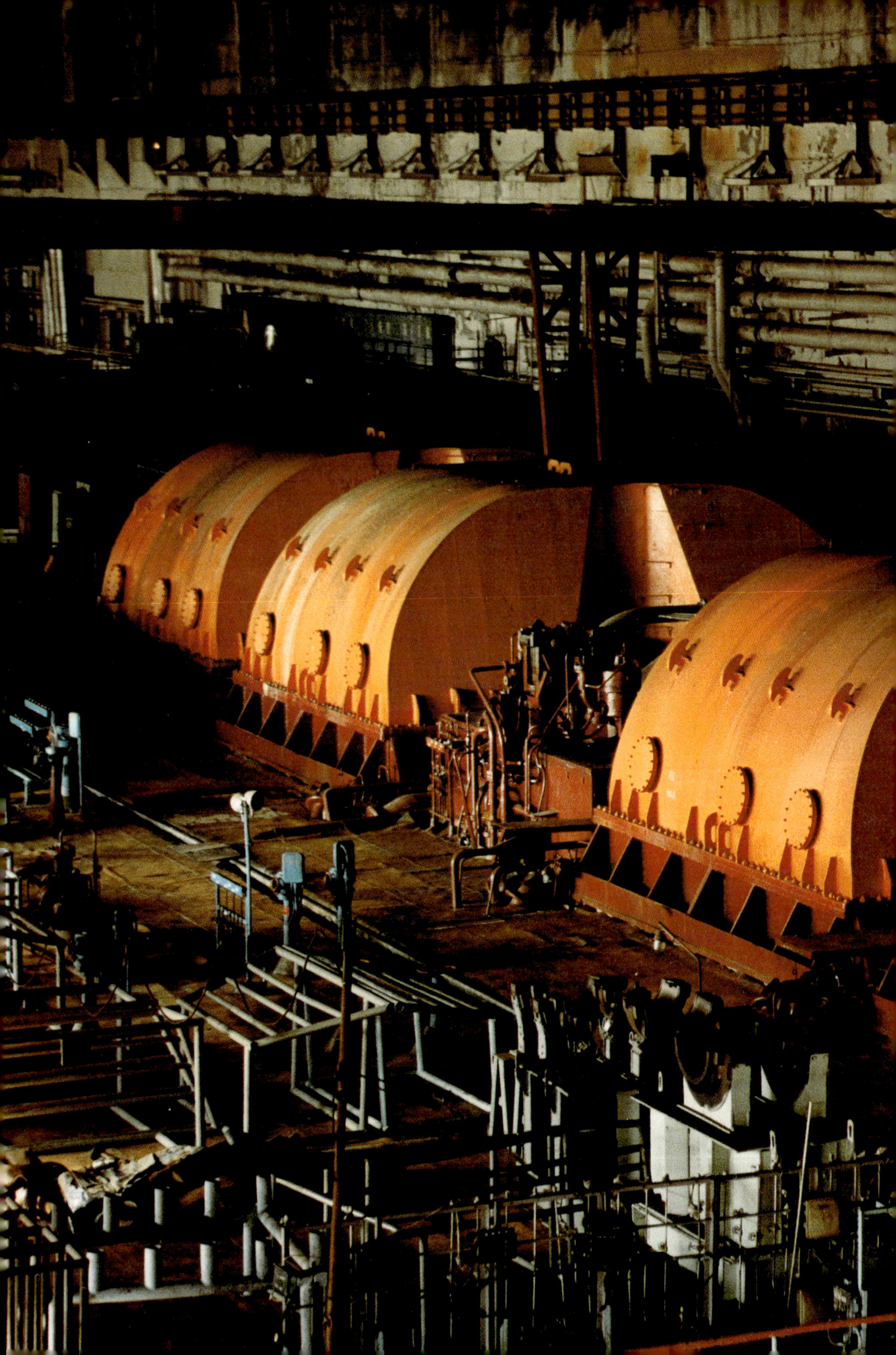

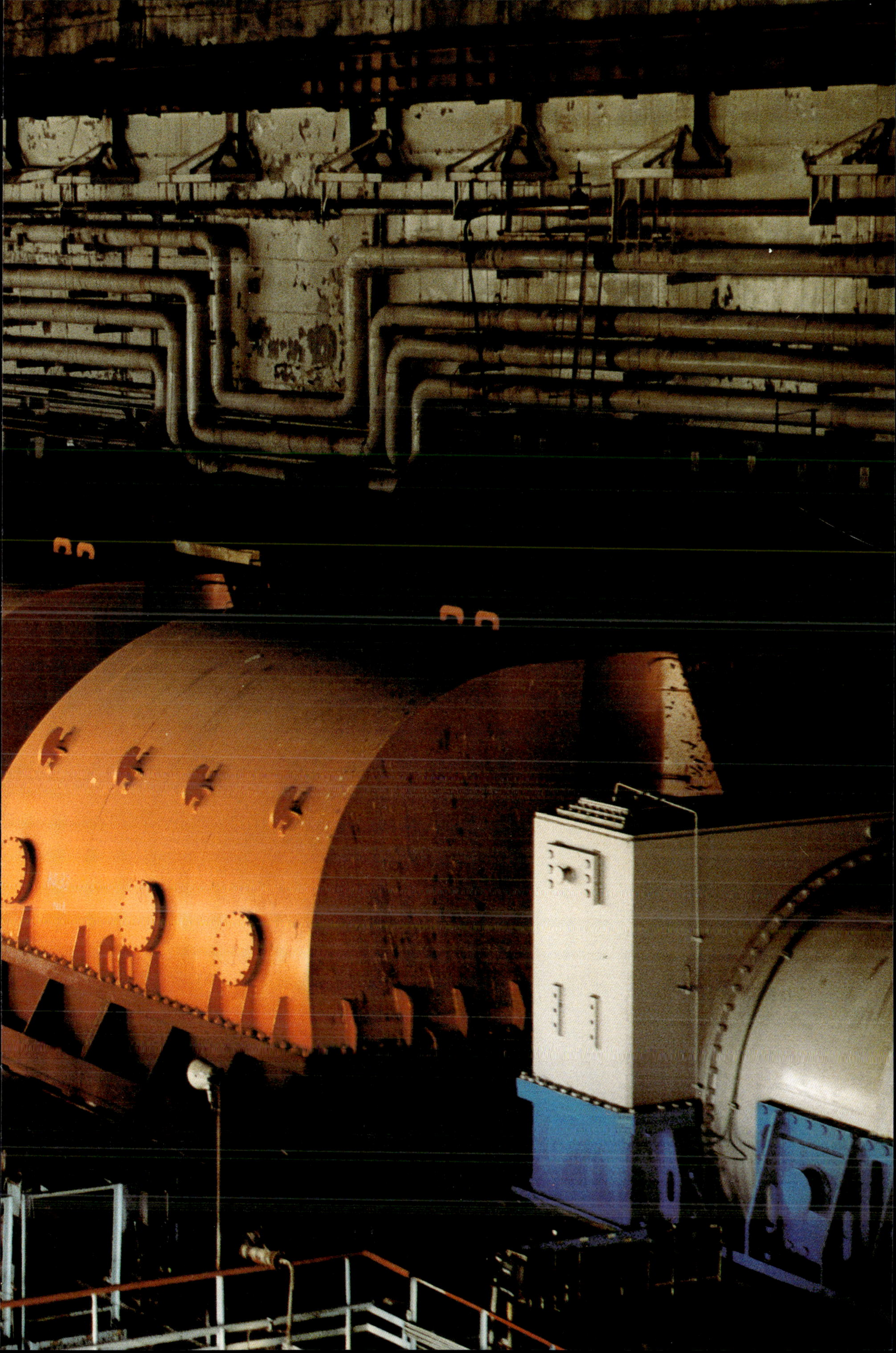

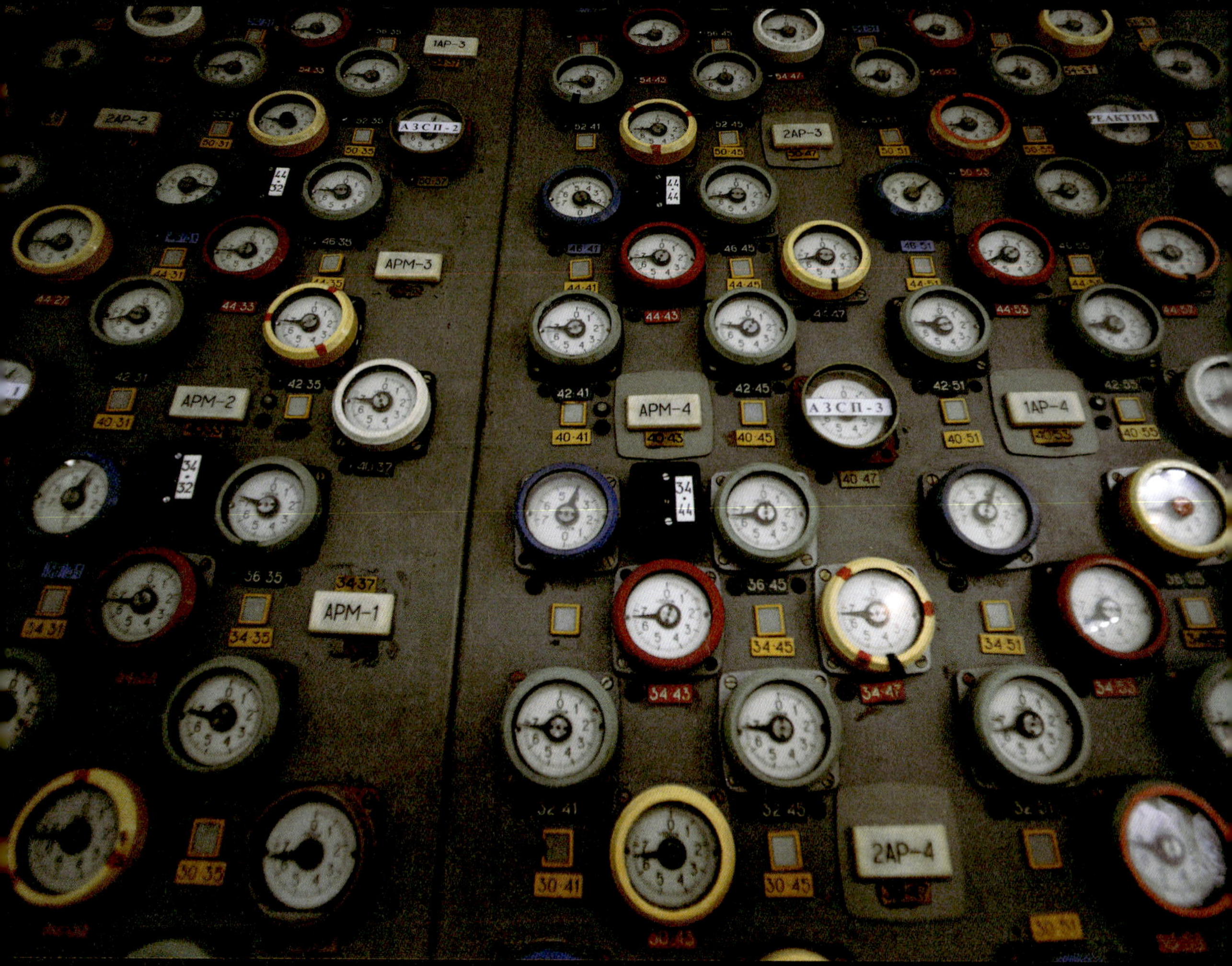
1АР-3
2АР-2
АЗСП-2
АРМ-3
АРМ-2
АРМ-1
2АР-3
АРМ-4
АЗСП-3
1АР-4
2АР-4

壁 の 向 こ う に ……

　通訳のアレクセイと僕は汚染防護のための白装束をまとい、発電所の広報部員に引き連れられて天井の低い薄暗い廊下を歩き続けた。終着の見えないリノリウム張りの廊下は3人の靴音を増幅させ、それが反響することでどこか闇の彼方に吸い込まれるような錯覚を覚える。

　どれだけ歩いただろうか。デッドエンドは十字架が刻まれた白い壁だ。いまだに行方が知れない作業員への弔いの花束が添えられたこの壁の向こうには22年経った今でも人間を秒殺できる高濃度の放射能という悪魔が生息している。悪魔の館4号炉に向かう禁断の鉄扉は閉ざされ、放射線警告標識が薄気味悪い笑みを浮かべながら睨みを効かしている。僕が行けるのはどうやらここまでのようだ。

　ここで働いていた人々は、肉体が危険に晒されていることなど教えられるはずもなく、平均よりも高い賃金と最新の暮らしと将来の地位の確約に明るい未来を想像していたはずである。長い廊下の先には光明があったのである。原子力の理論は地球が安全に存在するための根本のパワーバランスを危険に晒す代わりに、人間のみが必要とする莫大なエネルギーを得るシステムである。しかしひとたびそのシステムが支障をきたすと、僕らの先に見えるのは闇ばかりである。今この悪魔の館を封じ込めるために、再び莫大な地球の財産が使われようとしている。このあまりの代償の大きさ。

　悪魔の館を覆う石の棺桶の周りには野生のノハラジカを始め、発電所稼働時には姿を消していた様々な野生動物が暮らし始めているという。人間にとっての闇が野生にとっての光明になったのか？　この大きな石の棺桶は地球上での僕たち人間の存在意義を問う試金石でもあるかもしれない。

Beyond The Wall

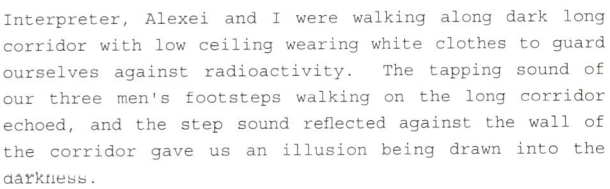

Interpreter, Alexei and I were walking along dark long corridor with low ceiling wearing white clothes to guard ourselves against radioactivity. The tapping sound of our three men's footsteps walking on the long corridor echoed, and the step sound reflected against the wall of the corridor gave us an illusion being drawn into the darkness.
How long did we walk? We came to the dead-end where there was a white wall carved the Cross on it. Beyond this wall, even after 22 years lapse, Devil called radio- lives and can kill human-beings instantly. In front of the wall a bunch of flowers was offered to console the soul of the lost workers of the power plant. The way leading to No 4 reactor was shut with heavy iron door and a radioactivity warning poster was put on it. It was as if gazing at us with a weird smile. It was a dead-end, and we were not allowed to go on further.
People who had been working here were not noticed having been threatened their lives at all. They must have dreamed of their bright future coming up with the prospects for drawing high wages, getting modern life style and high social standing. There they ought to have a bright future beyond their long hardship. Atomic theory is the means for only mankind to be able to get enormous quantity of energy in place of jeopardizing the basic power balance needed to keep the earth safe. But once this system breaks, we are sure to be thrown into pitch dark. Now a great sum of money is going to spend to seal up this facility with concrete. What a huge sum of cost!
There around 'Devil's Mansion' being covered with concrete, which looks like a huge stone-coffin, lot kinds of wild animals and plants begin to herd. Darkness for human-beings may become a paradise for wild creatures. This huge stone coffin may be a touchstone which asks our existence on the earth.

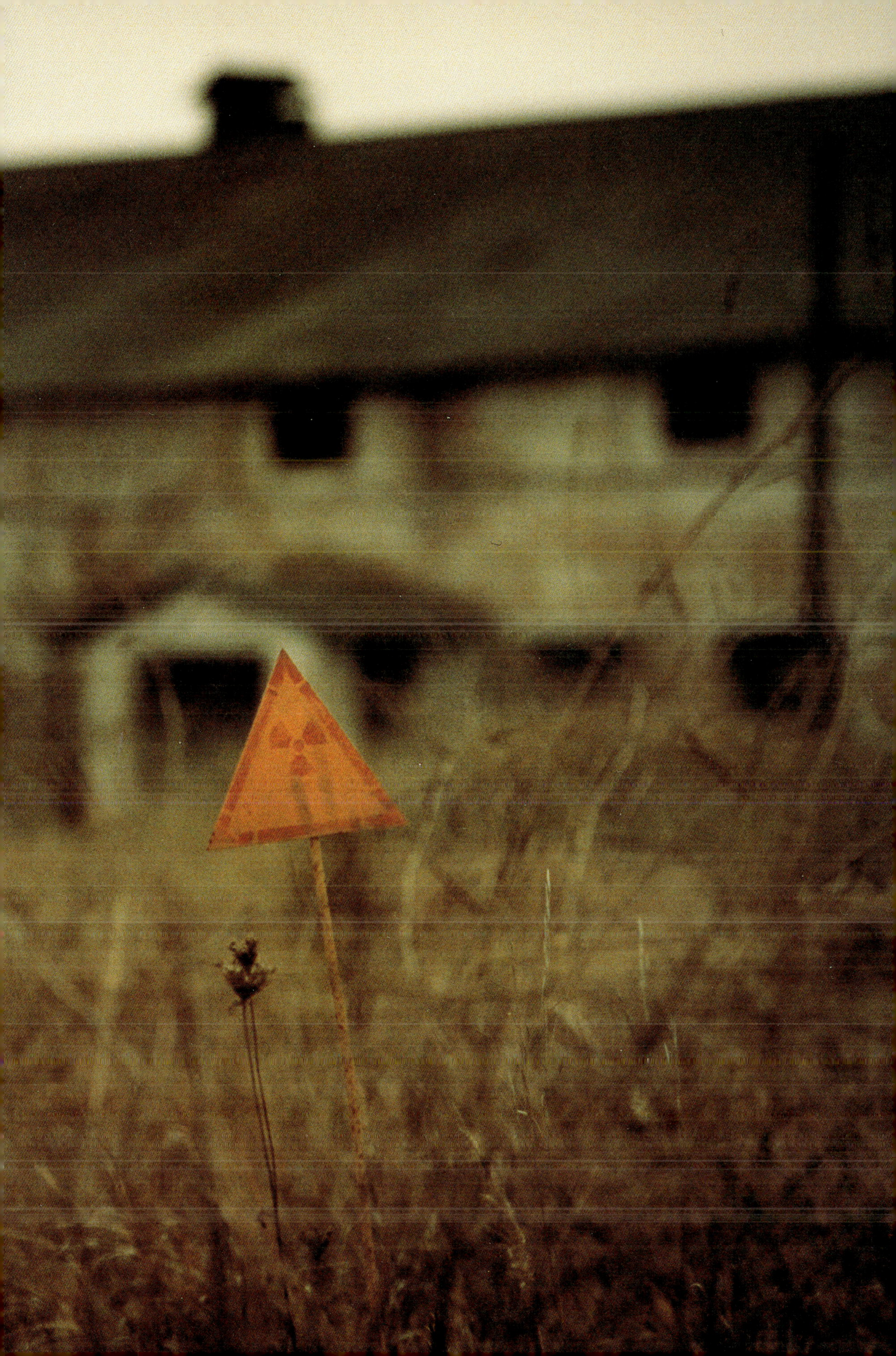

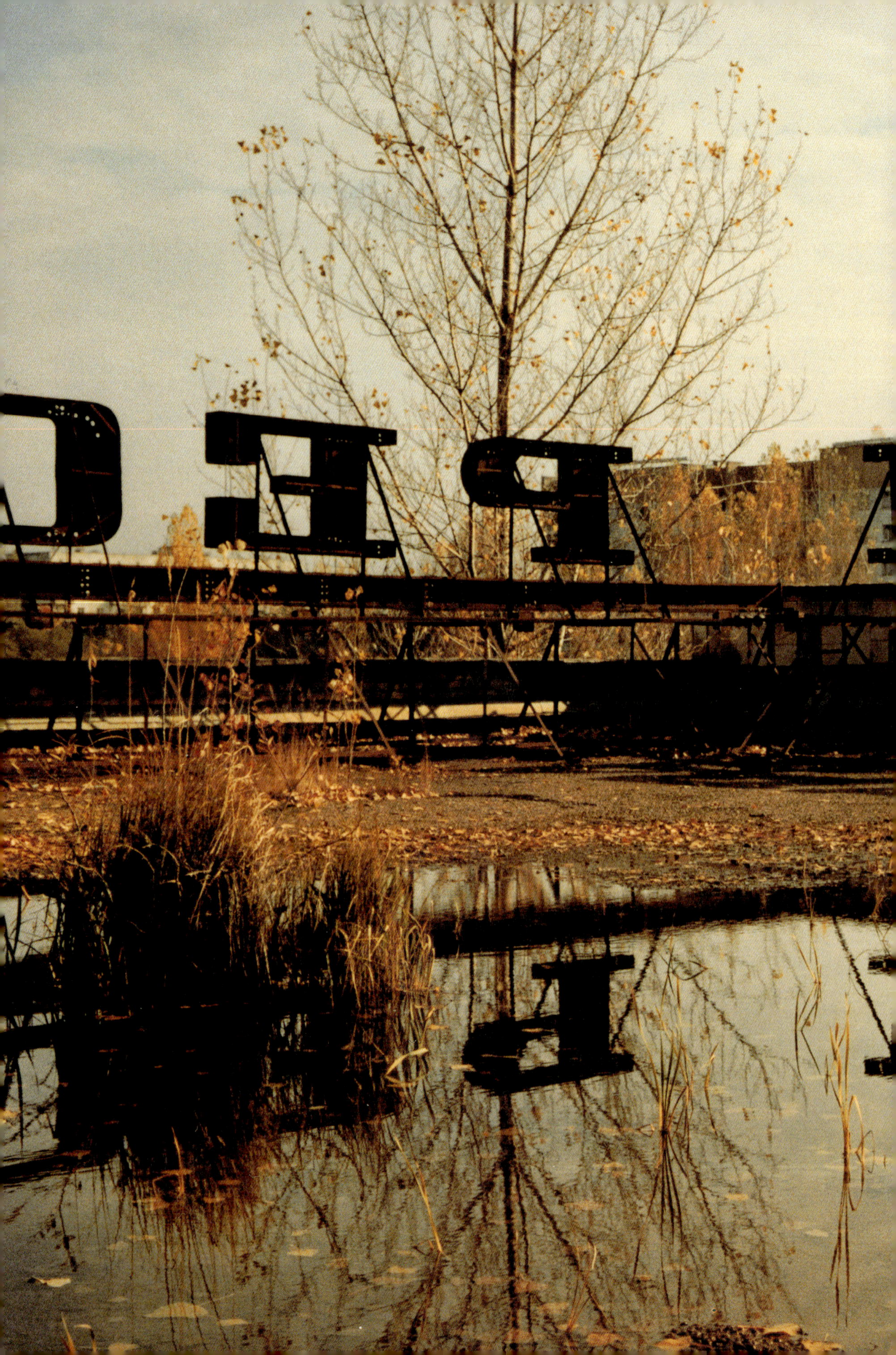

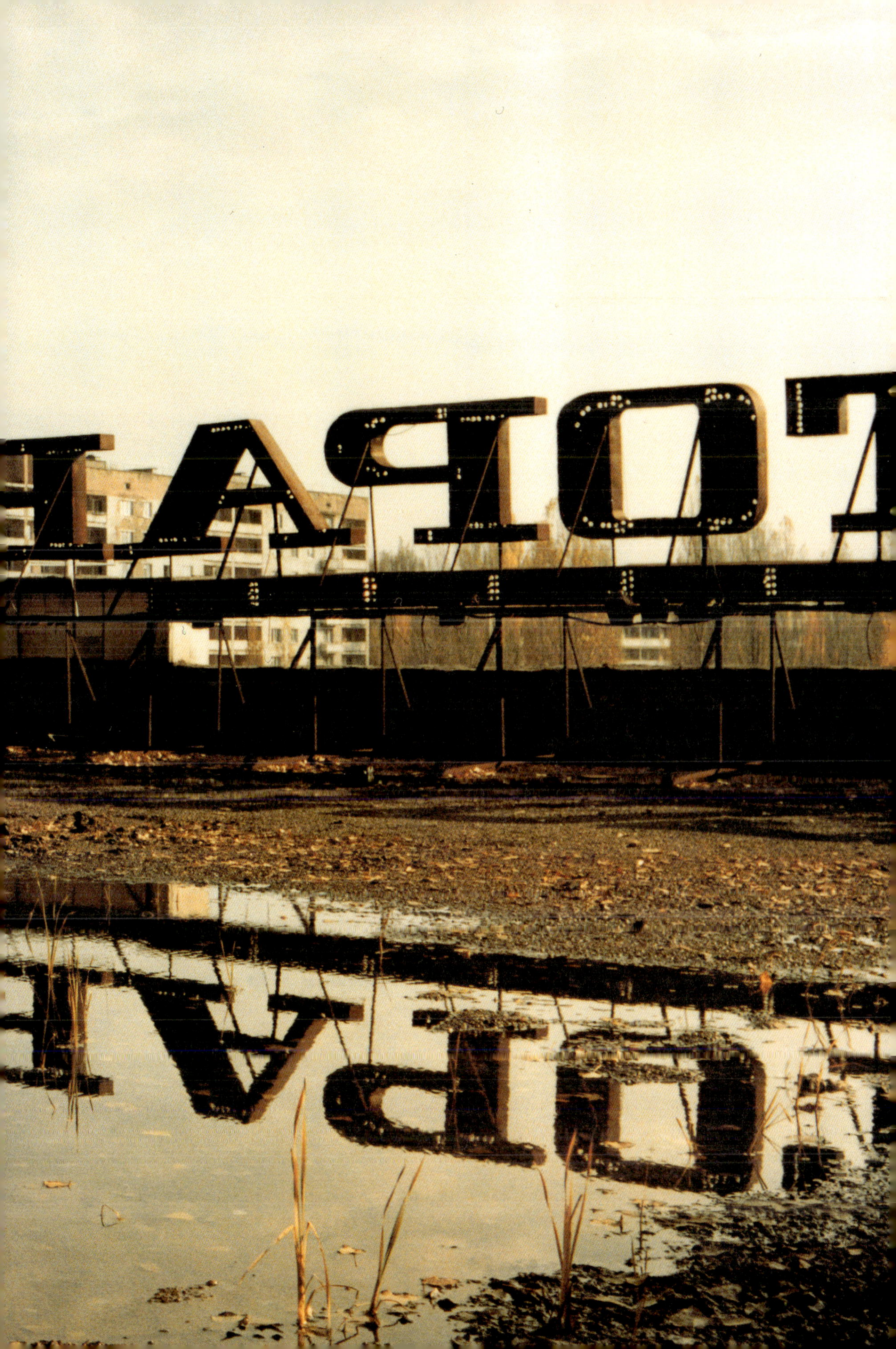

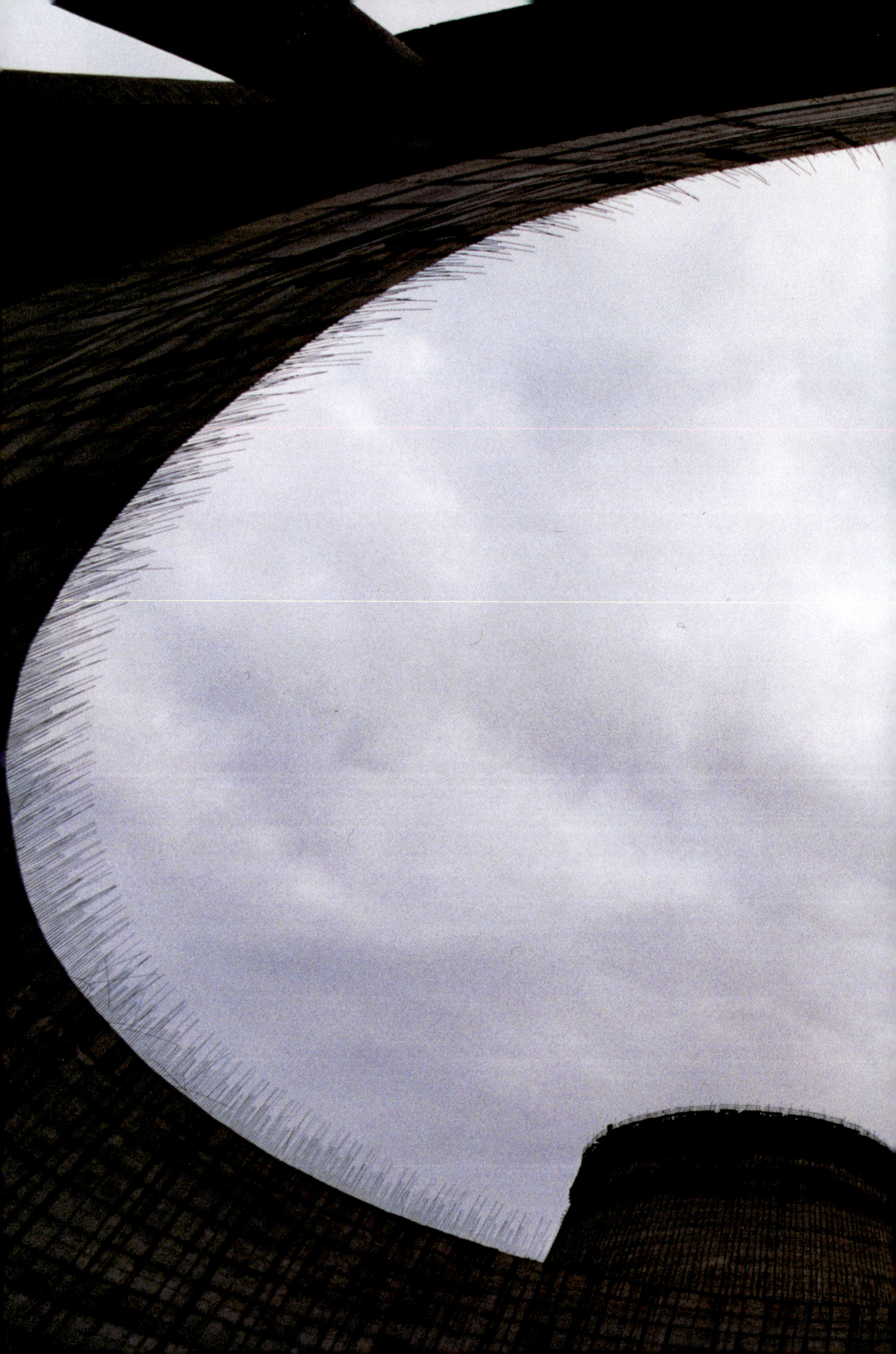

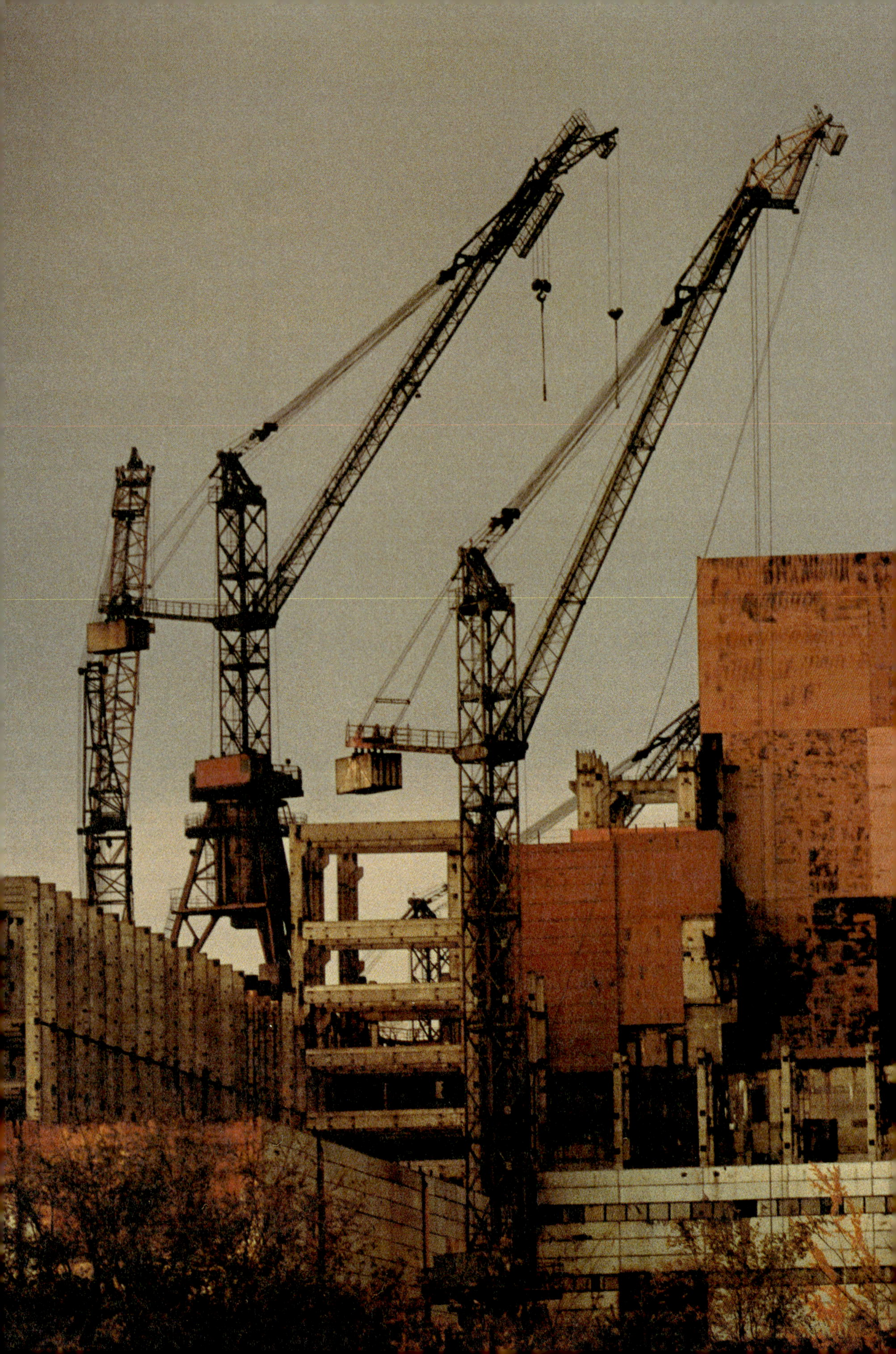

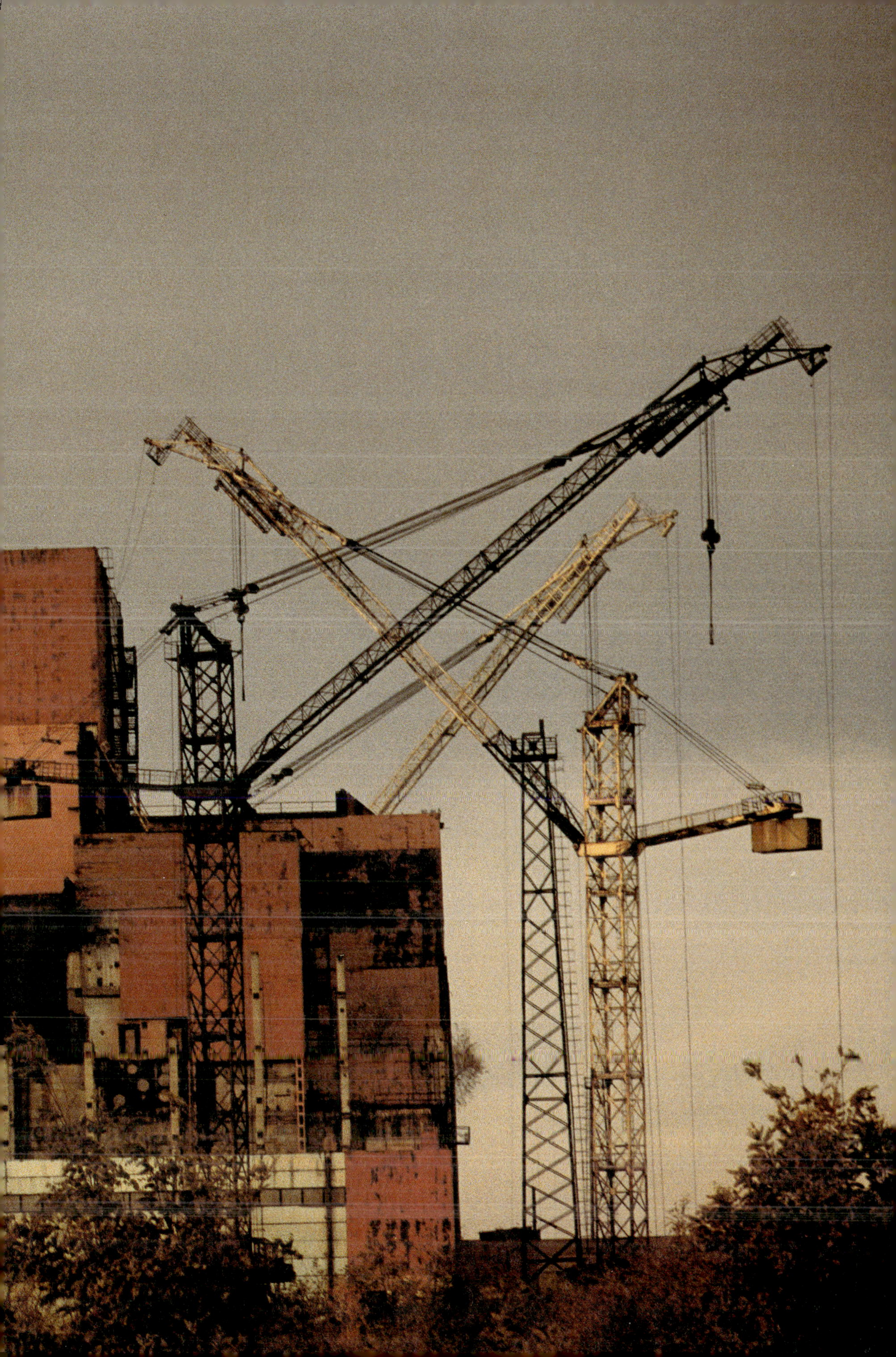

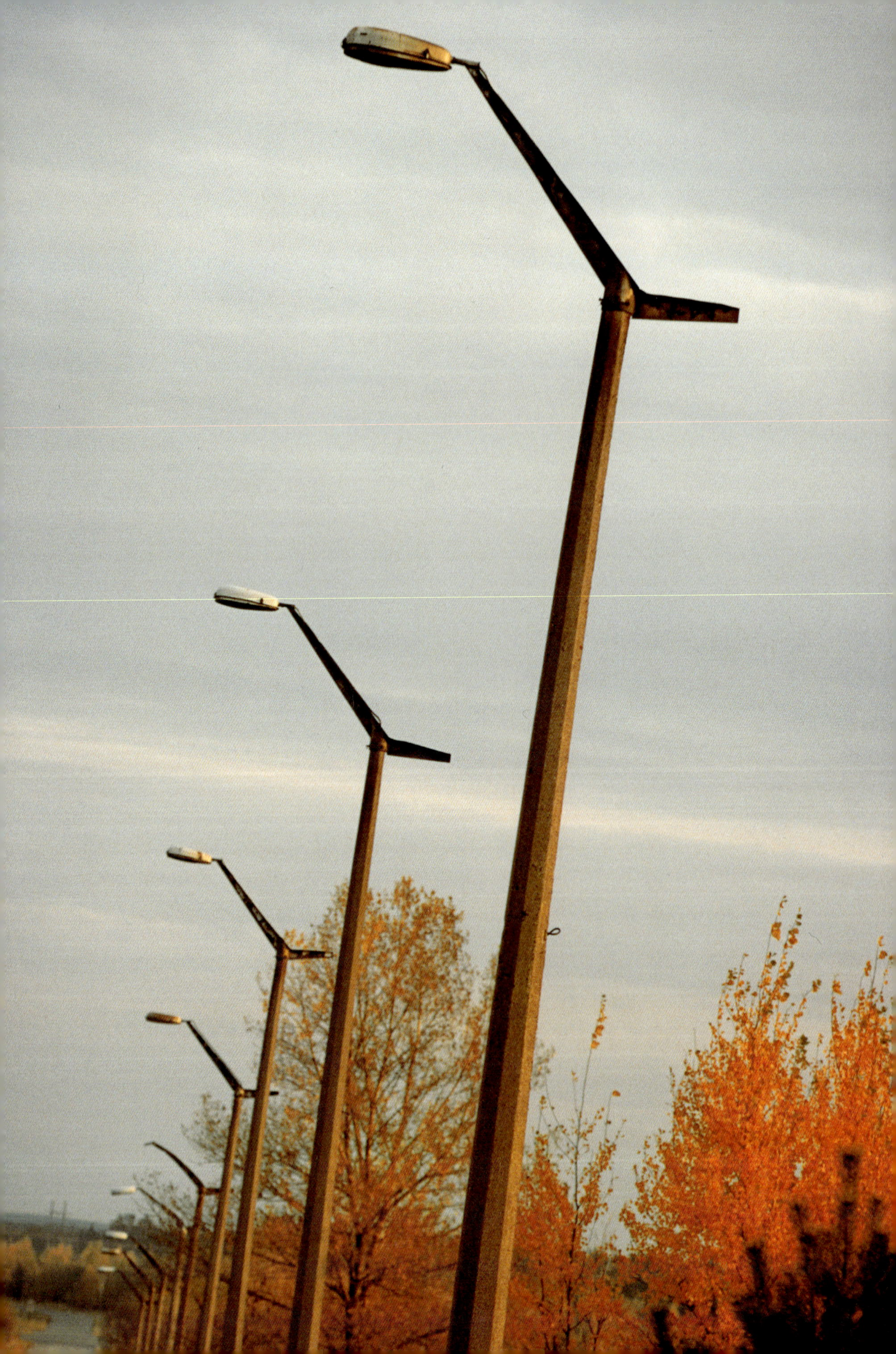

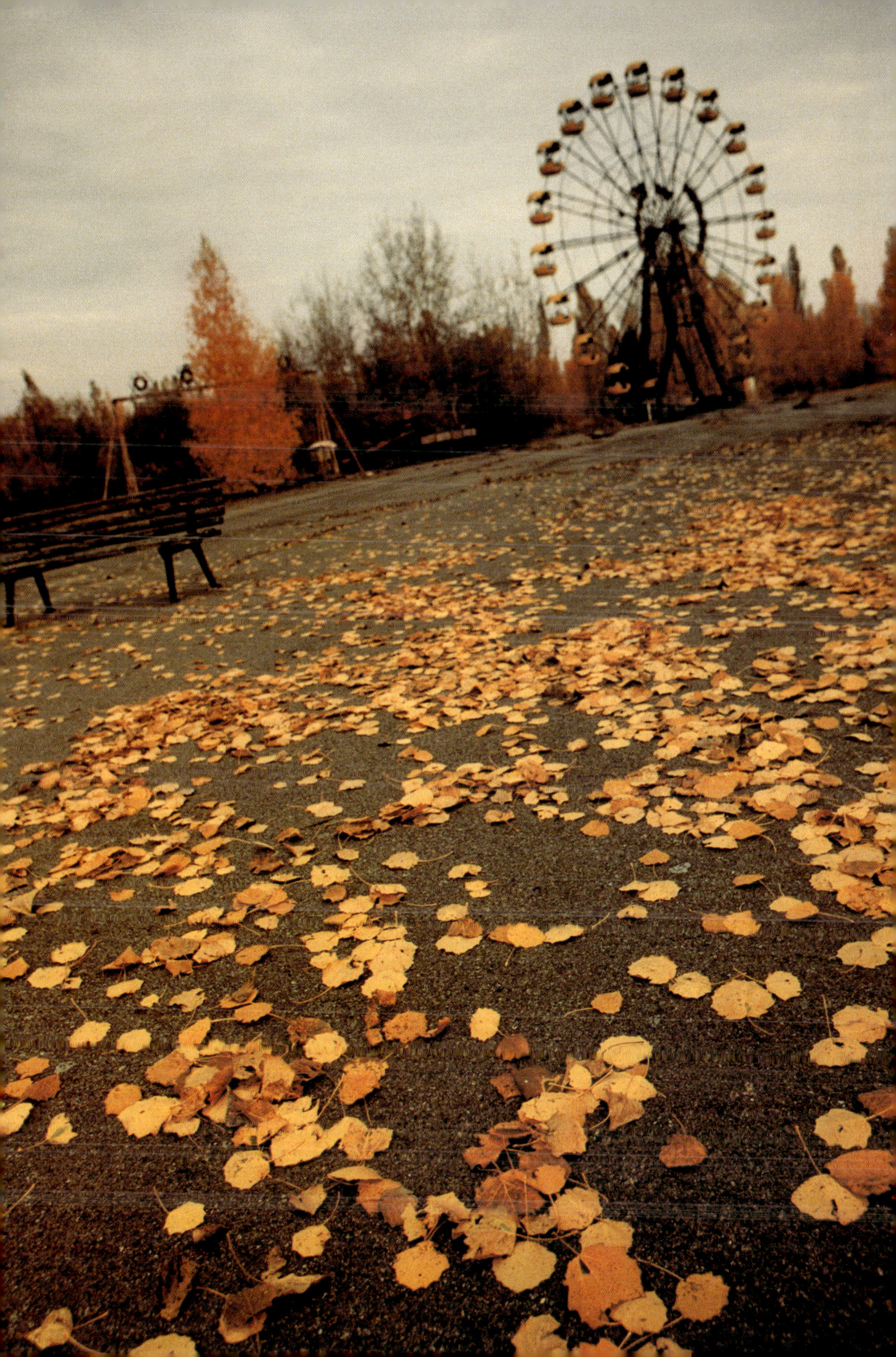

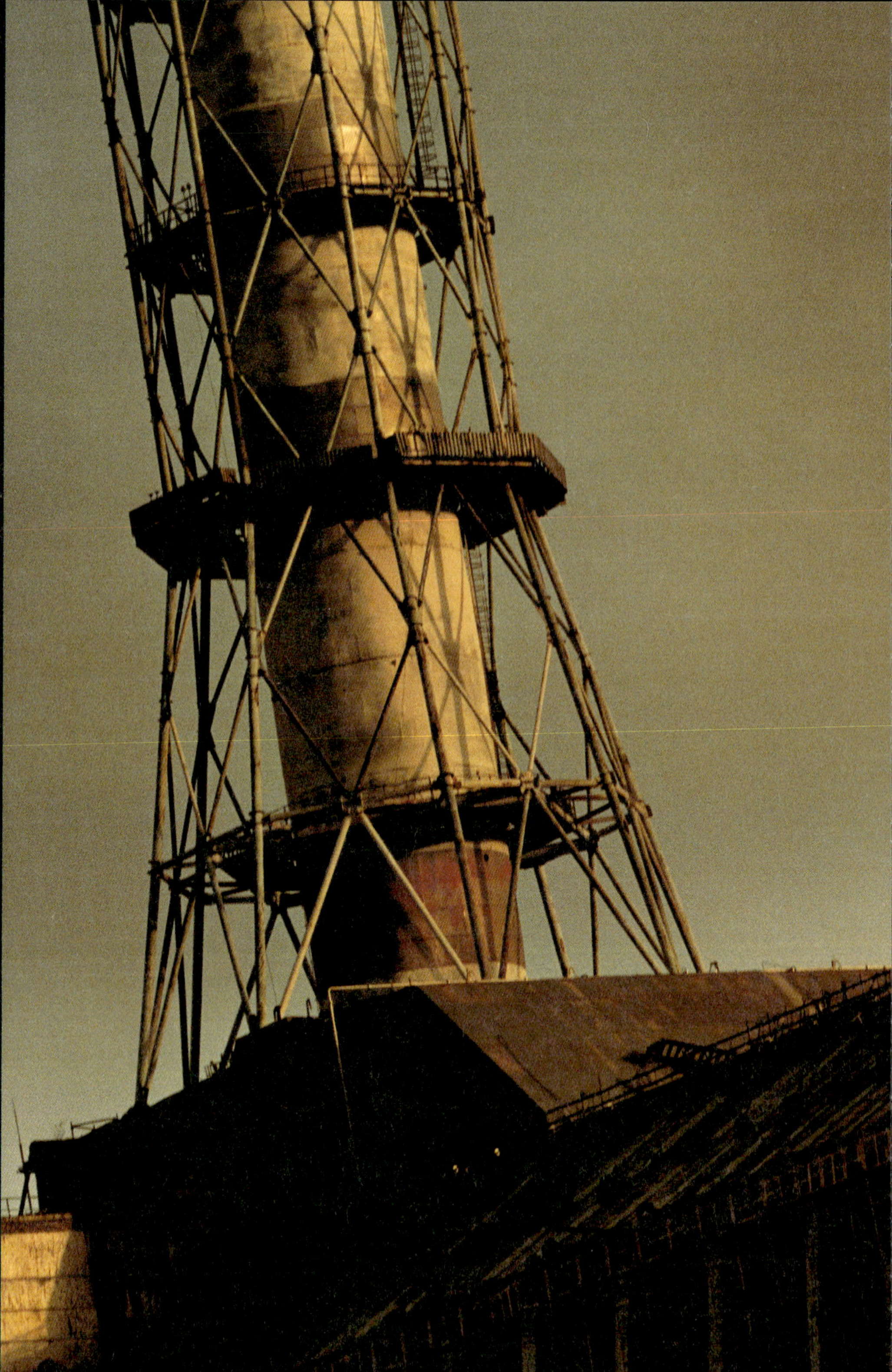

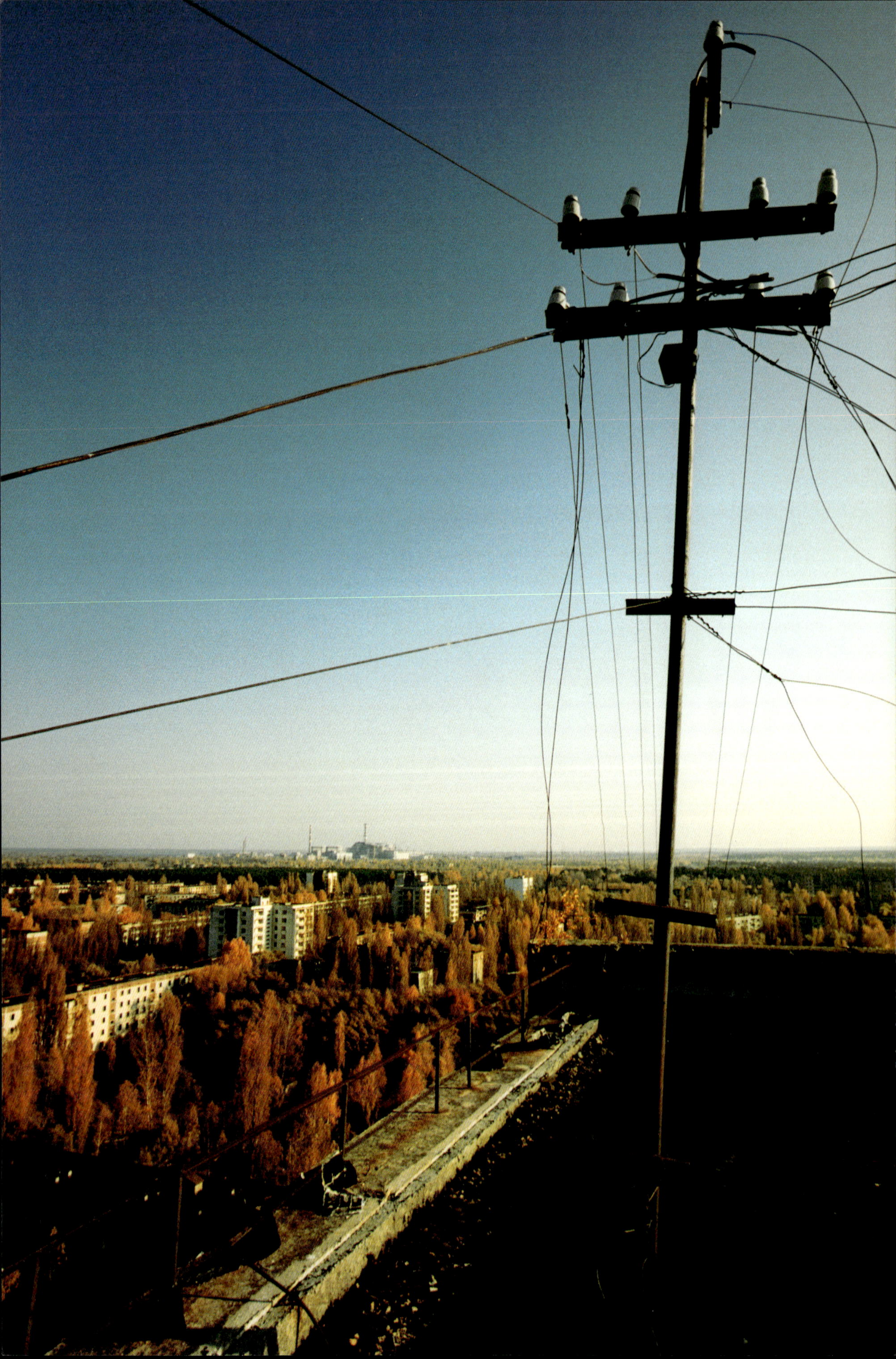

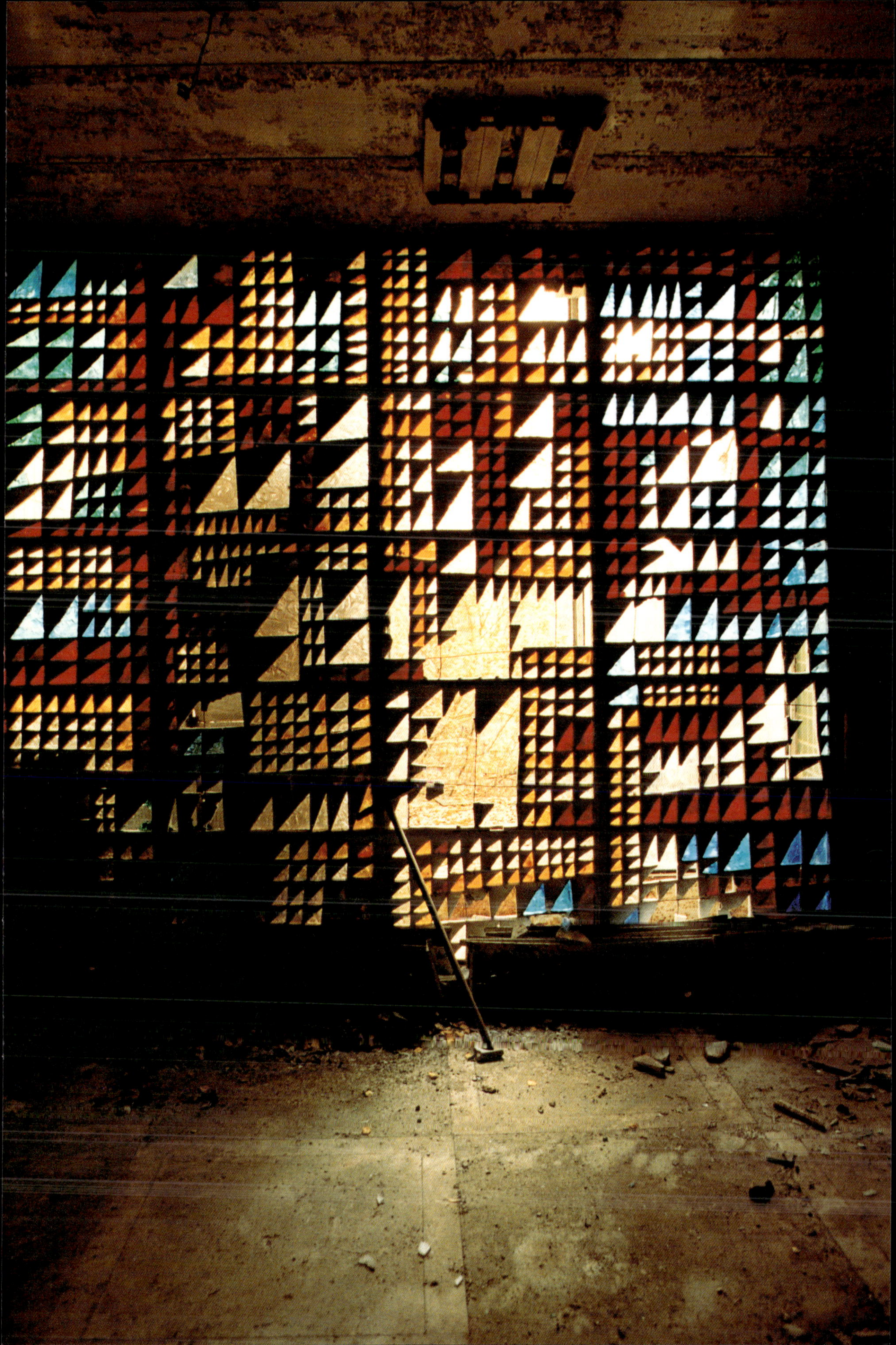

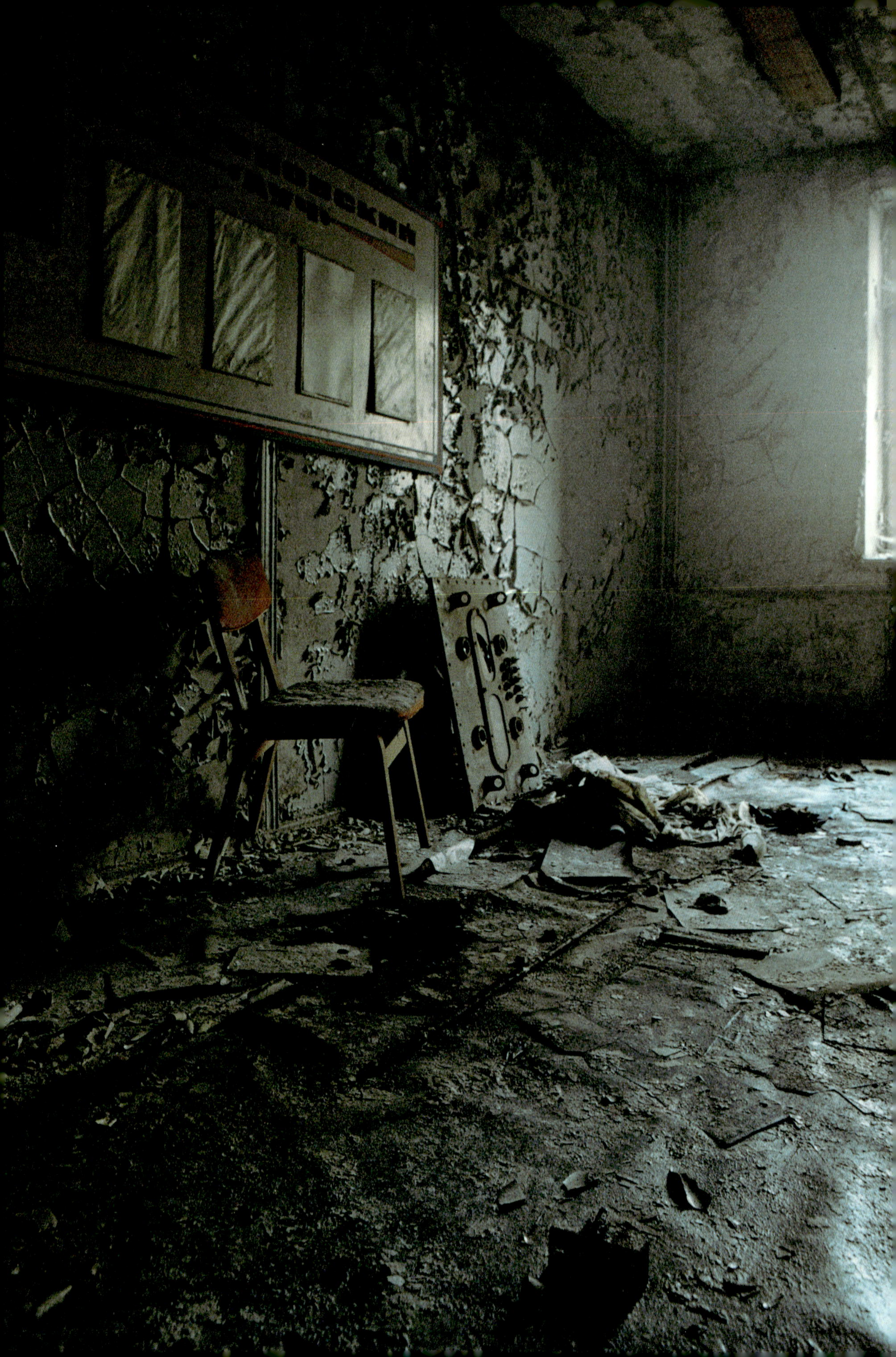

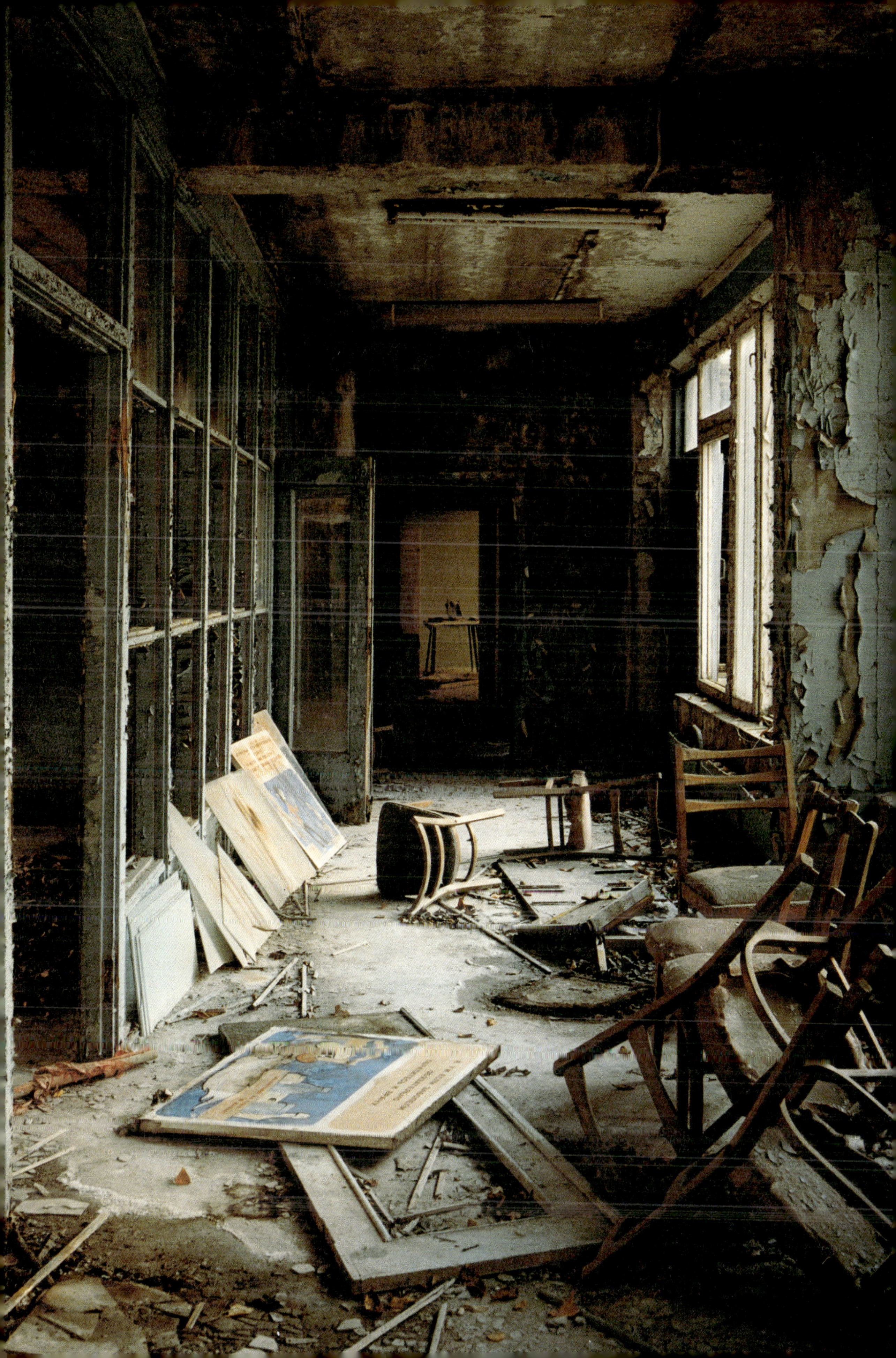

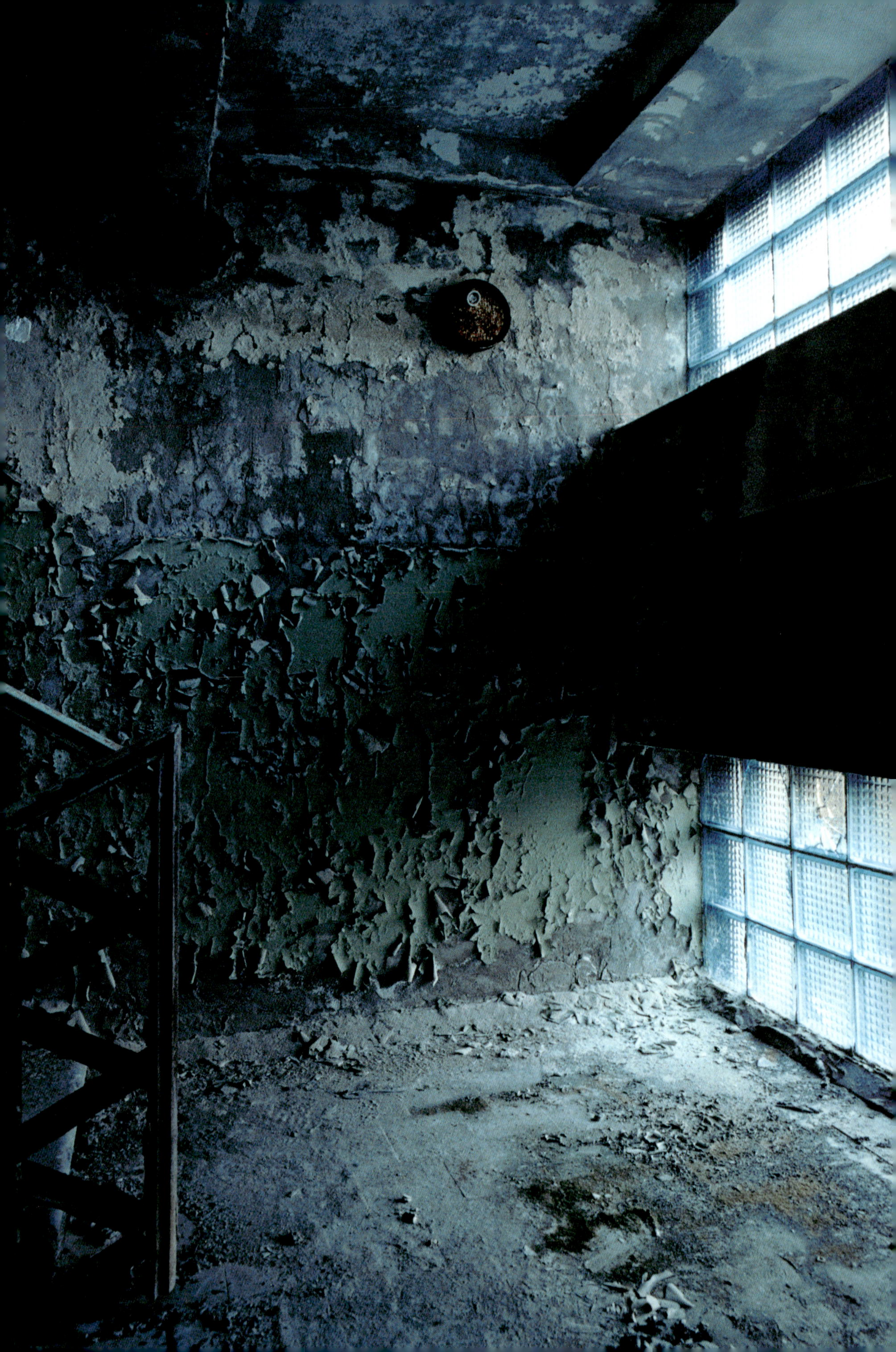

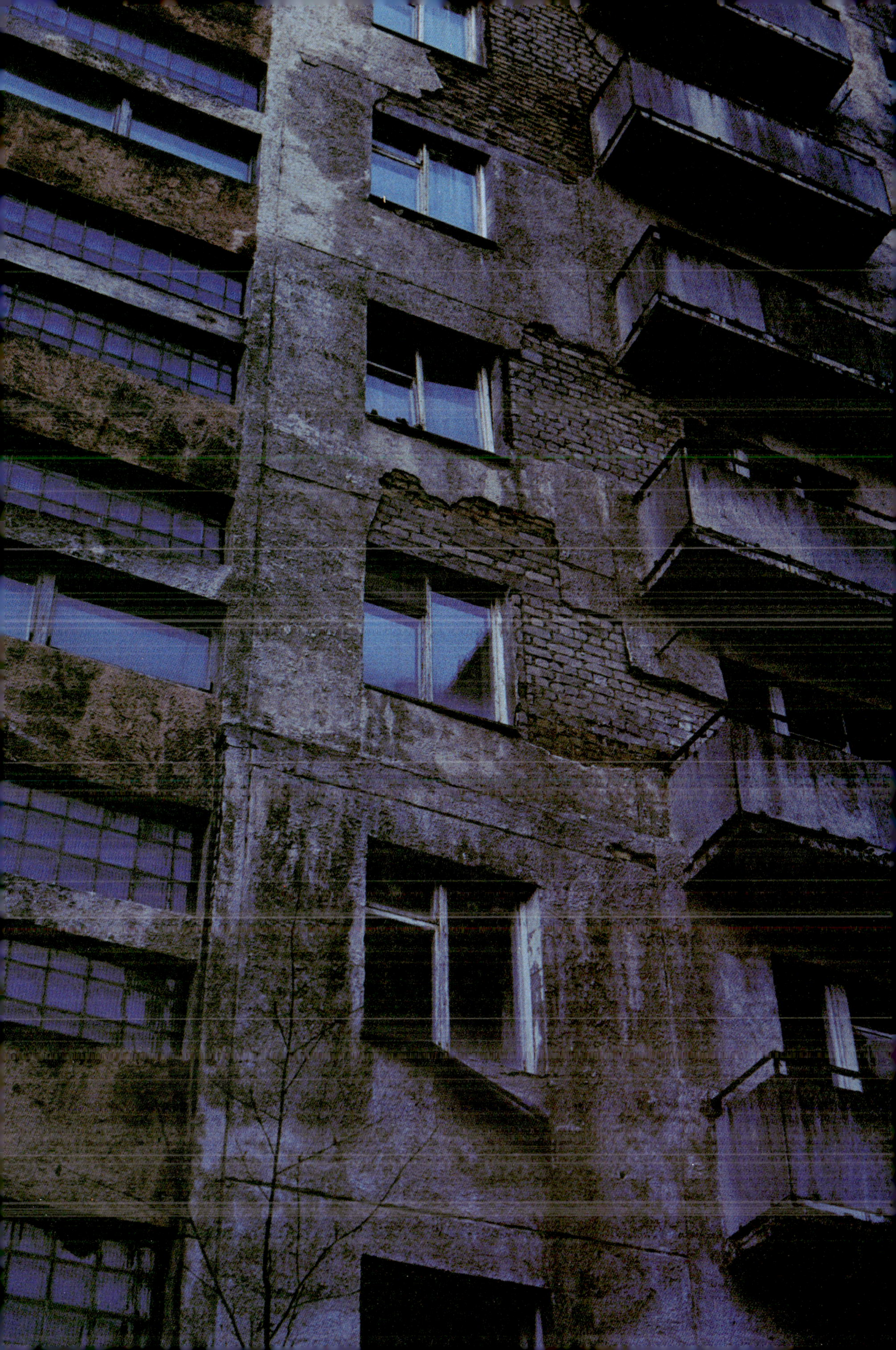

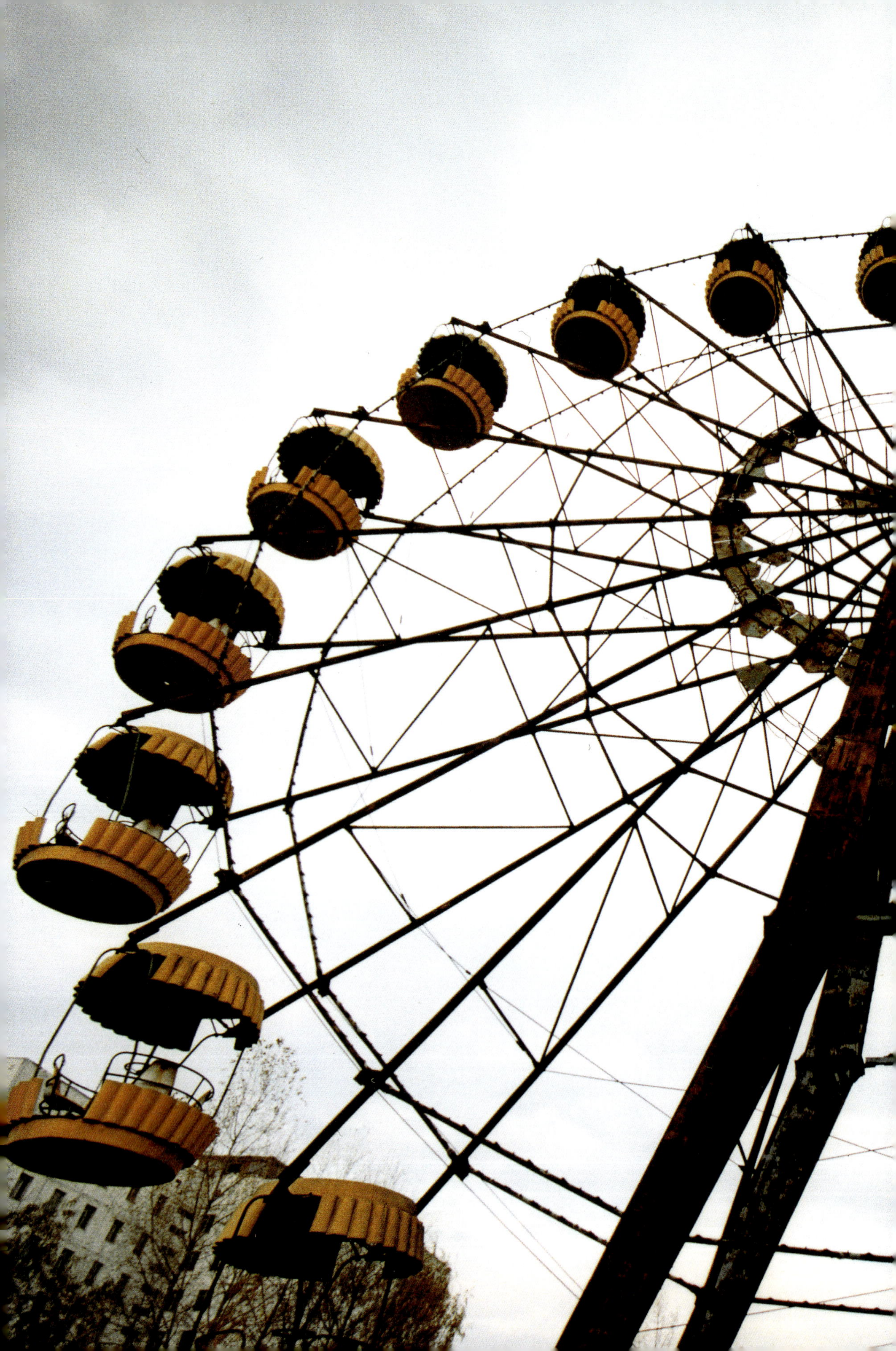

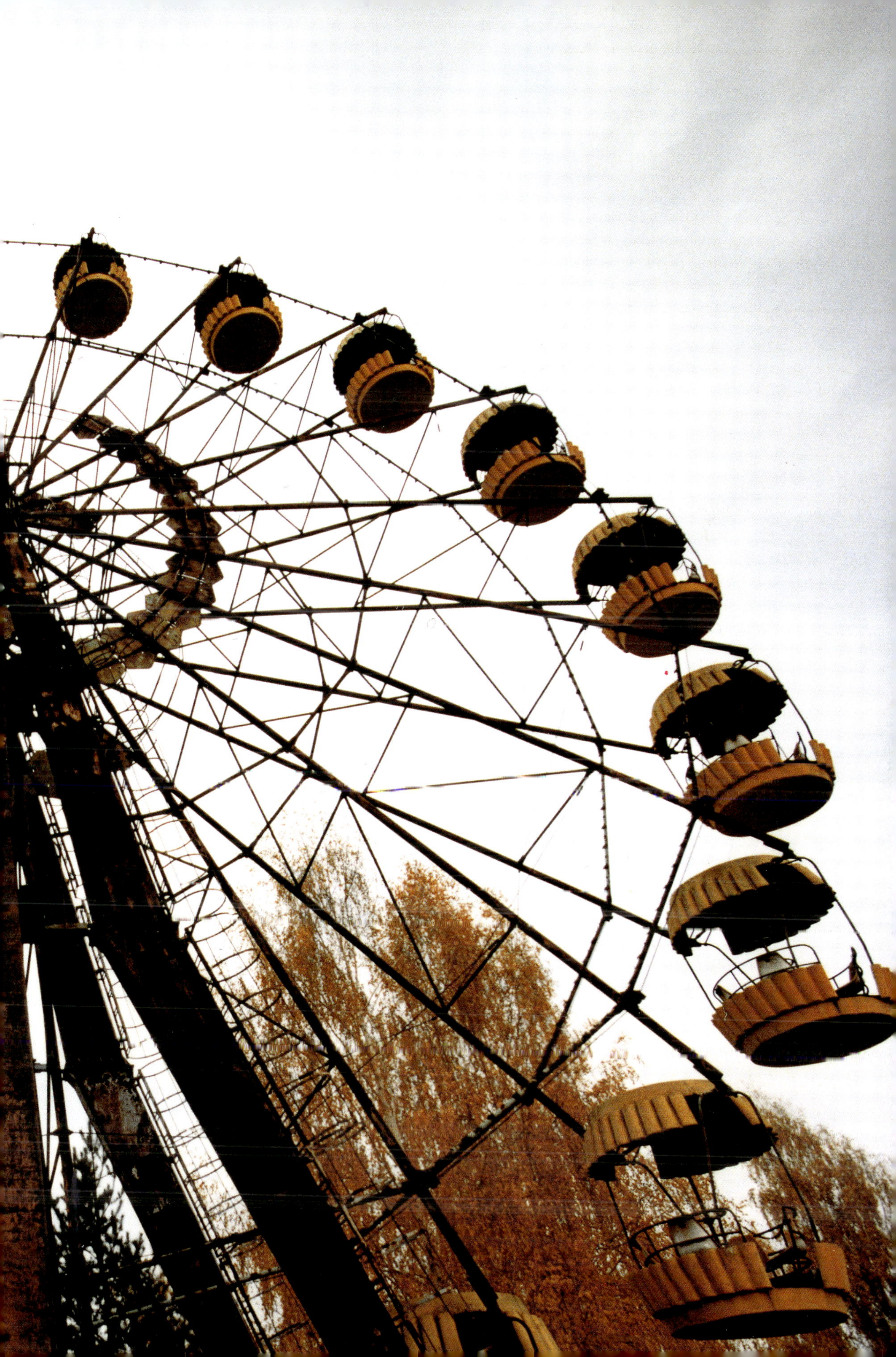

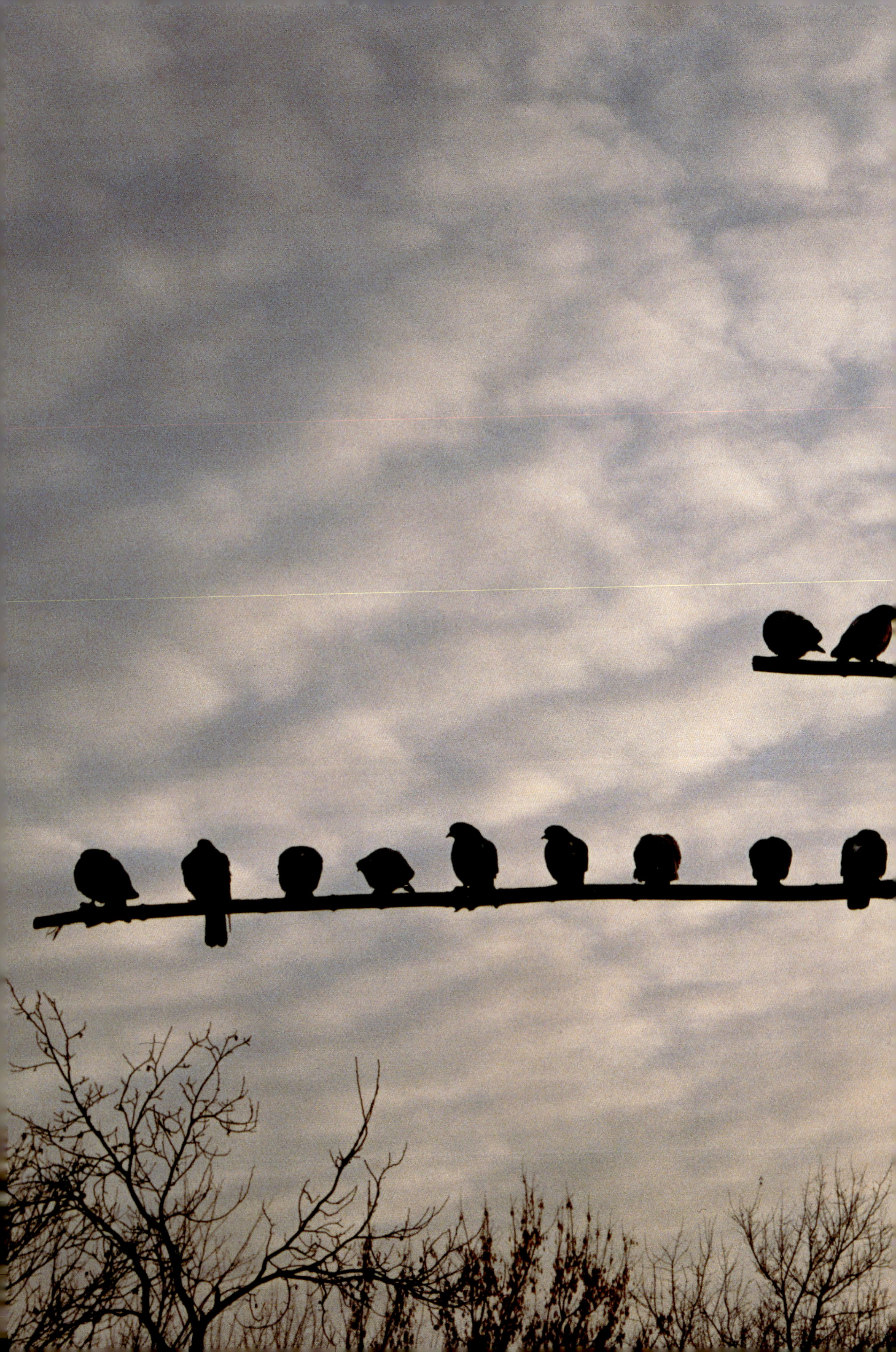

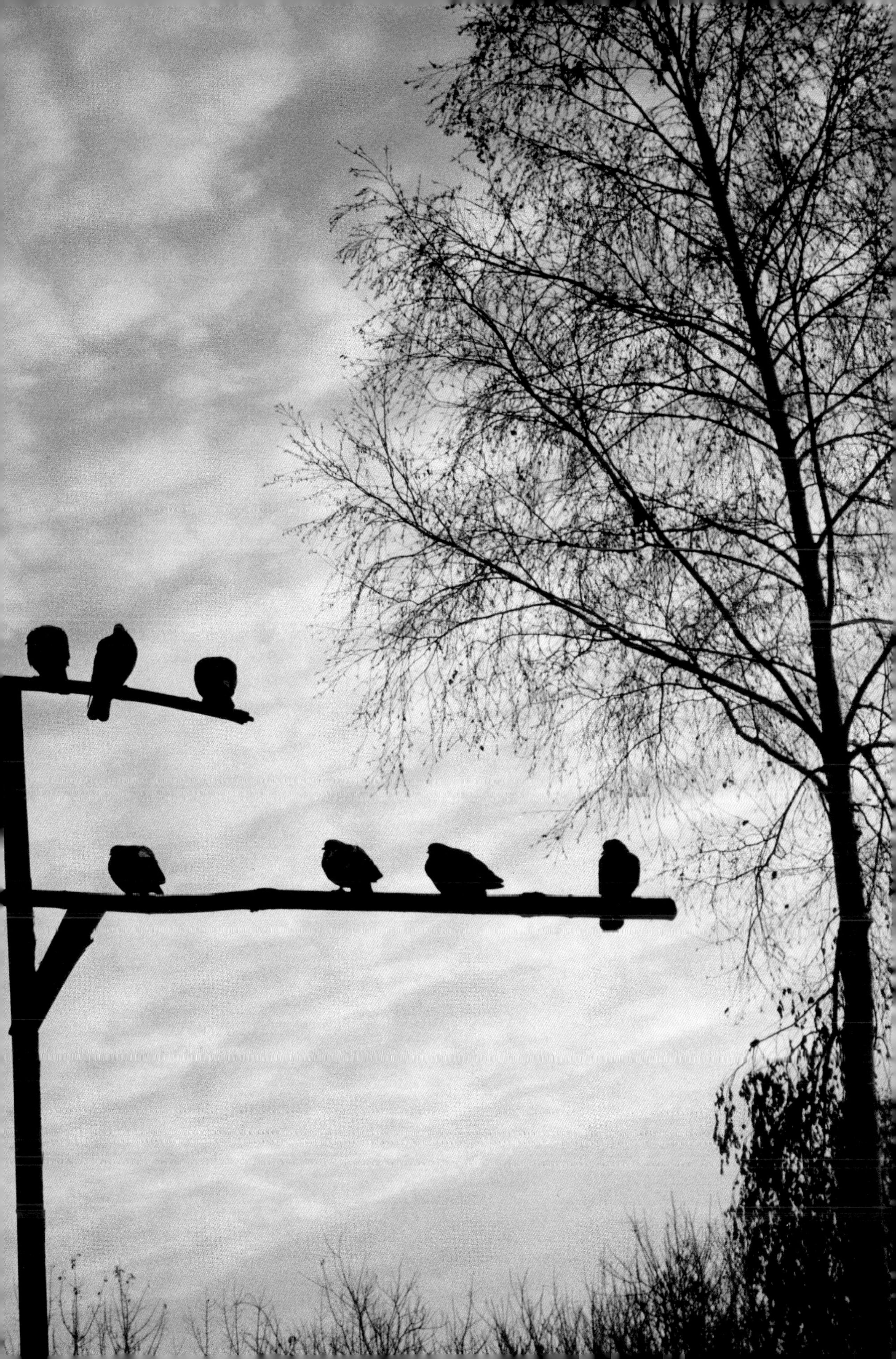

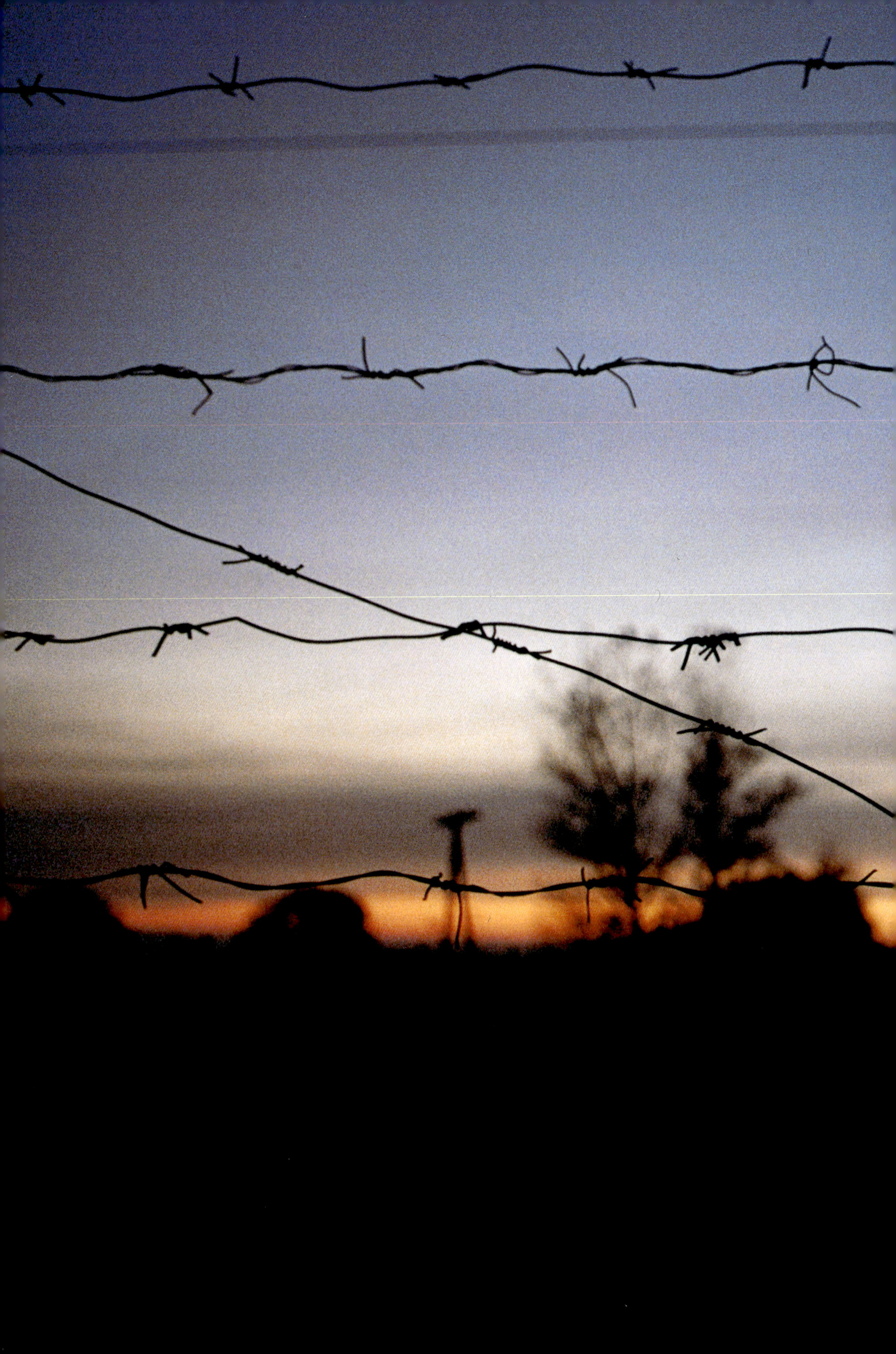

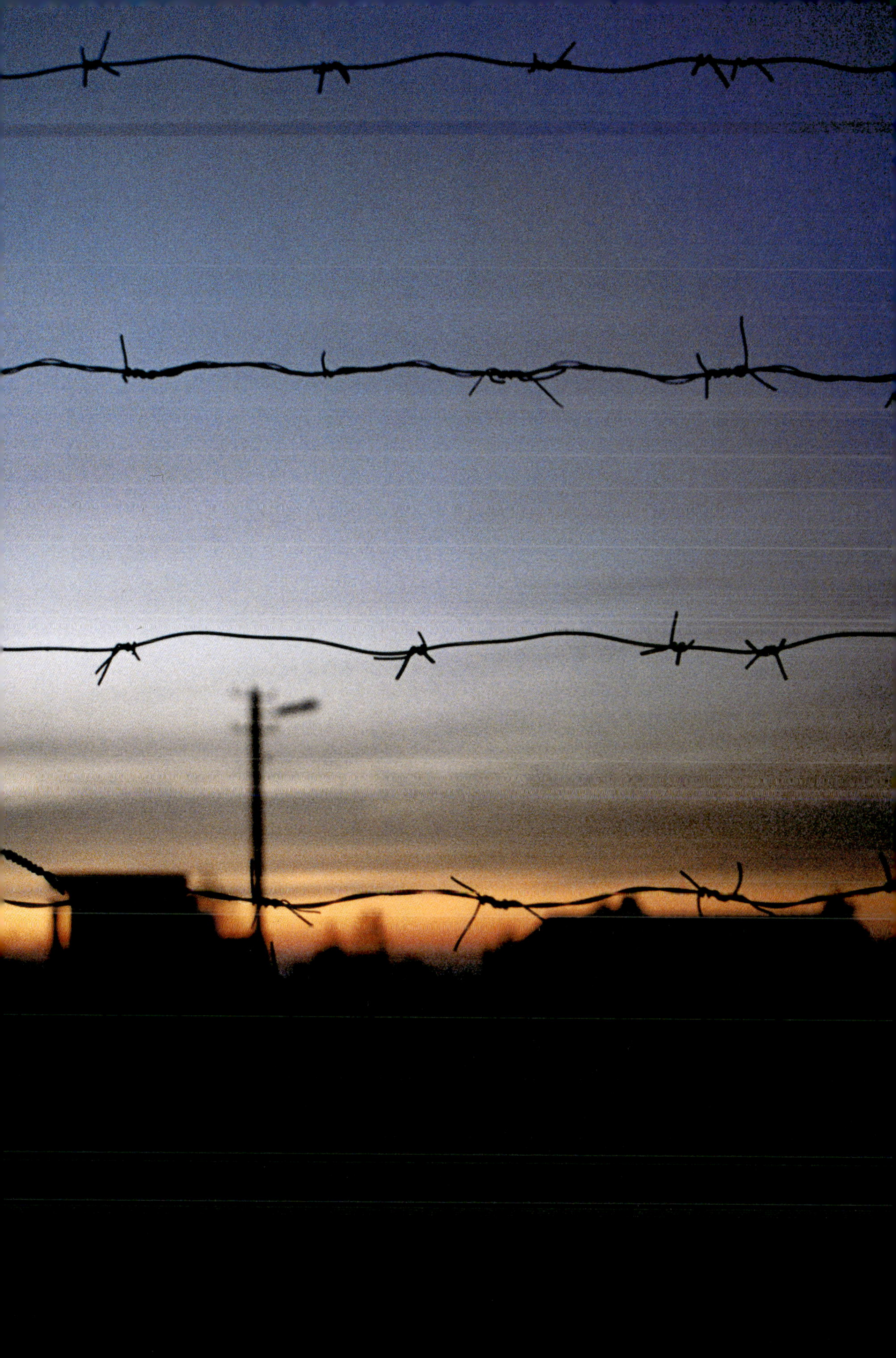

あとがき

　写真というのは見えないものを写す作業だと思っている。1枚のプリントから無限の世界が広がって、見る人すべての頭の中におのおのの小説なりドキュメンタリーができる。そんな不思議な力を持っていると信じている。
　僕が長年もの間廃墟を撮り続けている訳は、写真の持つそんな力をより明確にしてくれる高貴な存在だからかもしれない。ファインダー越しに覗いた廃墟の断片には季節の光も、建物の歴史も、こもった空気のにおいも、悲喜こもごもの人間の表情も写り込んでいるような気がしてならない。廃墟で三脚を構えファインダーをのぞいていると写真を撮ってるのではなく撮らせていただいてるような感覚に陥るのはそのためかもしれない。
　チェルノブイリの廃墟撮影も同じ気分だったのはいうまでもない。世界を震撼させた発電所事故から22年、いまだに居住はおろか入域も規制されているその地域は、世界に類を見ない核汚染された廃墟として時の流れを受け止めていた。見えない放射能の恐怖が常に意識にあり、ガイガーカウンターの電子音に肝を冷やす撮影旅行であったが、原発に関してノンポリだった僕があまりにも凄まじい現実を目の当たりにさせられたことで、改めて核について考えるいい契機にもなった。まずはチェルノブイリの核廃墟に手を合わそう。
　そして難題の多い辺境の撮影旅行をこちらのわがままを聞き入れながらうまくアレンジしてくれたウクライナ人スタッフ、通訳アレクセイ氏、案内人サーシャ氏、ドライバーセルゲイ氏に感謝。みんなその道のプロ、いい仕事してくれた、ありがとう。また現地レポートの記事掲載を快諾してくれたミリオン出版の中園努氏、高瀬元志氏、宮市徹氏、また「ナージャの村」撮影スタッフでもあるドキュメンタリーカメラマンの山田武典氏には様々な助言をいただき感謝。素敵な装丁とデザインをしてくださったヤマシタツトム氏、短い期間で翻訳を仕上げてくれた宮崎剛氏、不思議なプロフ写真を撮ってくれた藤原聡子女史にもこの場をお借りして感謝。そして何より、なぜか打ち合わせにマスクをして登場する二見書房の米田郷之氏のフットワークよい出版の実現、写真の編集、構成などの尽力に感謝。最後に、身勝手な海外撮影を励まし許してくれた僕の家族にありがとう。

2008年　4月26日　チェルノブイリ事故22周年　　　　　　　　　　　　　　　　　　　　　　　　　　　中筋純

Postscript

I believe that photograph is a kind of product which will figure out the invisible things. Infinite space will be developed from a sheet of print, and it will allow everybody to compose novel and documentary.
The reason why I have been taking pictures of ruins for many years is that unknown power which photograph has may seem to be something noble to me. When I take picture I always feel that bright light of season, history of building, stuffy smell of air and mixed emotions are all confined in the view which I am going to select through viewfinder. At ruins whenever I look through a viewfinder to take picture, I feel I am allowed to take picture not to take. That is due to the above mentioned strange fantasy I felt.
Not to say I had the same feeling when I visited Chernobyl. 22 years after the terrible accident which threatened the whole world even now this area is forbidden to enter not to pay of immigration. This fact tells of the seriousness of the damage this city suffered. During this snap tour, I had been threatened by indivisible radioactivity and warning sound of Geiger counter. Though I had been indifferent about nuclear power plant, I came to realize the importance of antinuclear weapon observing serious damage of the power plant accident. Anyway, I offered my condolence to the victims who were killed by the accident. And I would like to express my deepest gratitude to the Ukrainian staffs; interpreter Alexei, guide Sasha, driver Sergei, who arranged this tour granting my selfish demands. They were all professionals. You all have done a good job! Thank you! I would like to say my hearty thanks to all people. I appreciate your many kindnesses shown to me during this tour; Mr. Nakazono, Mr. Takase, Mr. Miyaichi in Million Publishing Co.,, who willingly accepted to find a space for this on-the-spot-report, documentary cameraman Mr. Yamada who gave me a lot of precious advices, Mr. Yamashita who gave me a beautiful book design, Mr. Miyazaki who translated the report into English in a short time, Ms. Fujiwara who offered me fantastic professional pictures. Above all, I appreciate Mr. Yoneda in Futami Publishing Co.,. His good favors given me on cutting and composition of my report. By the grace of his efforts quick publication has been realized. Last, I would like to give thanks to my family.

April 26, 2008
On the 22nd memorial year of Chernobyl nuclear plant accident
Nakasuji Jun

```
Index
Sheet-1
```

001 コパチ村国営農場付近

002 チェルノブイリ原子力発電所。左奥が事故のあった4号炉

003 プリピアチの高層アパートより発電所を望む

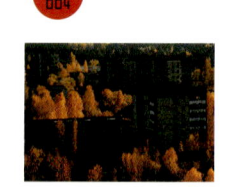

004 暮れなずむアパート群。街路樹が黄金色に輝く

005 ソ連共産党幹部専用のマンション壁面。

006 アパート屋上。たまった雨水は建物を浸食する

007 町を覆う鉄条網越しに見るアパート群。

008 プリピアチは1970年に建設。地図に載らない秘密都市だった

009 文化会館前の放射能汚染されたリンゴ

010 プリピアチのランドマーク16階建て高層アパート壁面

011 学校壁面。ソ連の教育は小中一貫だった

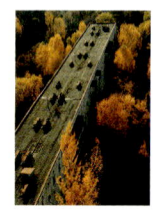

012 ポプラに埋もれる低層アパート

013 16階高層アパートより遊園地を望む

014 観覧車近景

015 観覧車近景

016 1度も人を乗せることがなかった観覧車

017 遊園地にはゴーカートや空中ブランコもあった

018 プリピアチの墓地。3ミリレントゲン／時以上の高濃度汚染地帯

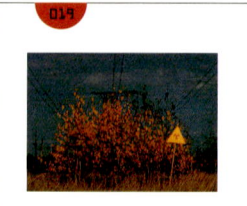

019 放射能警告標識。通称三ツ葉マーク

020 幼稚園内の寝室

Index
Sheet-2

021 果てしなく長い中央病院入院棟廊下

022 バスターミナルのロッカールーム

023 発電所幹部アパート内部。壁面には色味があるが、家具は質素だ

024 「チェルノブイリ」と並べられたロシア語教育用サイコロ

025 中央病院小児科病棟の1室

026 ファミリータイプのアパート内部。2DKが基本

027 ファミリータイプアパートの居間

028 家電製品は事故後のどさくさで盗難にあったものが多い

029 発電所幹部アパート内のキッチン

030 打ち捨てられた人形

031 独身アパートの共同部屋

032 朽ち果てていく音楽室の椅子とレコード

033 小中学校音楽室内部。音楽教育は盛んだったようだ

034 幼稚園遊戯室

035 幼稚園ロッカールーム

036 小中学校視聴覚室

037 市民プール内観。飛び込み台も併設

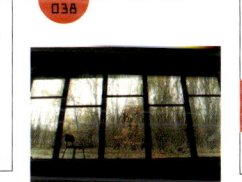

038 長距離バスターミナル待合室内部

039 小中学校教室。ガスマスクは事故後持ち込まれたものだそうだ

040 中央病院小児病棟。チェブラーシカはここにもいた

Index
Sheet-3

041 小中学校図書室。湿気で朽ち果てた書物
042 市民会館ホール。暗闇に浮かぶ弦の切れたグランドピアノ
043 市民生活センター内の理髪店
044 子供をモチーフにした影絵の落書きがプリピアチの至る所で見られる
045 中央病院歯科治療室の診察台
046 小中学校音楽室内部
047 小中学校音楽室の朽ち果てたレコード
048 中央病院小児病棟治療室の小型無影灯
049 小中学校美術室
050 幼稚園内部に打ち捨てられた人形
051 一般労働者アパートにもエレベーターは完備されていた
052 一般労働者アパート。朽ち果てたシーリングライト
053 幼稚園内部。時間が止まった空間
054 中央病院音楽ホール。瓦礫に埋もれそうなピアノ
055 一般労働者アパートの狭くて急な階段
056 小中学校体育館
057 核汚染の地の新しい原始の命
058 小中学校教室
059 小中学校玄関。ほころんだ赤旗
060 高等学校壁画。戦争讃歌

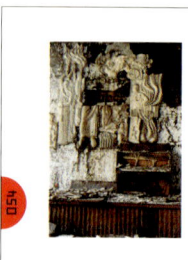

Index
Sheet-4

061 一般労働者アパート壁画。労働讃歌

062 一般労働者アパート壁画。原子力技術讃歌

063 一般労働者アパート。労働、教育讃歌

064 レーニン肖像

065 レーニン肖像

066 レーニン肖像

067 レーニン肖像

068 1986年メーデーの準備。市民会館倉庫

069 小中学校教室壁面。ソ連邦共和国国旗模型

070 小中学校教室

071 プリピアチ共産党事務所内倉庫に捨てられた国家紋章壁画

072 小中学校教室。生徒たちのノートが散乱

073 小中学校廊下。国家スローガンが書かれた壁新聞

074 レーニン肖像

075 市民生活センタースーパーマーケット支店のガラス扉

076 モスクワオリンピックのマスコット、ミーシャの看板

077 プリピアチ共同墓地付近。立ち入りが最も危険な場所のひとつ

078 チェルノブイリ市郊外の船の墓場。事故以来放置されている

079 事故復旧作業に参加したソ連軍戦車

080 船の墓場

Index
Sheet-5

- 081 プリピアチ消防署裏
- 082 チェルノブイリ市郊外の車両廃棄場
- 083 プリピアチ市内、工場従業員送迎バスの廃車
- 084 プリピアチ市内、工場従業員送迎バスの廃車
- 085 チェルノブイリ市郊外の車両廃棄場
- 086 プリピアチ市内工場。爆発時刻の10分後に停止した時計
- 087 コパチ村付近の放射線警告標識
- 088 チェルノブイリ発電所より原野を貫く高圧電線。遠くハンガリーまで送電
- 089 発電所付近。おびただしい数の高圧電線と原子炉冷却水路
- 090 西日に萌える発電所。プリピアチの通称「死の橋」より遠望
- 091 発電所に向かうひび割れた道路
- 092 発電所内部。1号炉から4号炉まで続く薄暗く果てしなく長い廊下
- 093 2号炉タービン室
- 094 2号炉タービン室俯瞰
- 095 2号炉制御室、原子炉圧力計
- 096 4号炉を隔てる壁。十字架が刻まれたデッドエンド
- 097 4号炉に向かう鉄のドア。うす笑みを浮かべる警告標識
- 098 コパチ村国営農場付近も高濃度汚染地帯だ
- 099 プリピアチ市民センター屋上
- 100 建設途中だった冷却塔の残骸

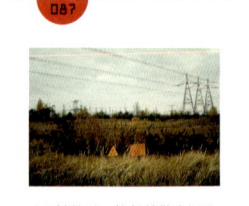

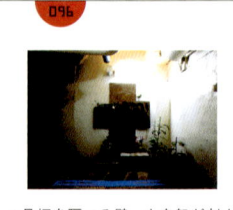

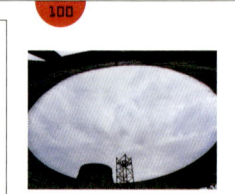

Index
Sheet-6

101 建設途中だった5、6号炉の残骸。22年間クレーンもそのままだ

102 プリピアチに向かう通称「死の橋」の街灯

103 枯れ葉舞う晩秋の遊園地

104 今なお不気味な輝きを見せる発電所煙突。手前が4号炉石棺

105 16階高層アパート屋上より発電所を遠望

106 文化会館の踊り場。ささやかな色味にほっとする

107 小中学校教室

108 小中学校給食室前の廊下

109 発電所幹部アパートの階段踊り場

110 夕刻の発電所幹部アパート外観

111 プリピアチのもうひとつのランドマーク。失楽園

112 人間不毛の地に他の生き物は静かに生き続ける

113 暮れなずむ30キロ立ち入り禁止区域の鉄条網

114 埋められた村コバチの放射線警告標識

115 車内でも鳴り響くガイガーカウンター

著者

中筋 純（なかすじ・じゅん）

1966年和歌山県生まれ。東京外国語大学中国語学科在籍中より、海外を放浪し、独学で写真技術を習得。卒業後出版社勤務を経て中筋写真事務所設立。ストリートファッション雑誌、アパレル広告をメインに、映画スチール、舞台広告、CDジャケット撮影など幅広いジャンルで活躍。廃墟撮影はライフワークでもあり著書に『廃墟探訪』（弊社刊）、『廃墟本1』『廃墟本2』（以上ミリオン出版）、『廃墟彷徨』（ぶんか社）、『廃墟、その光と影』『愛という廃墟』（東邦出版）がある（いずれも共著）。

廃墟チェルノブイリ
（はいきょ）

写真・文
中筋 純

発行所
株式会社　二見書房
東京都千代田区三崎町2-18-11
電話 03-3515-2311（営業）
　　 03-3515-2313（編集）
振替 00170-4-2639

印刷・製本
図書印刷株式会社

ブックデザイン
ヤマシタツトム

現地撮影コーディネート
アレクセイ・タナシエンコ
http://www.japanese-page.kiev.ua/jpn/index.htm

参考文献
『チェルノブイリの森　事故後20周年の自然誌』（メアリー・マイシオ著／中尾ゆかり訳／日本放送出版協会）
『チェルノブイリの真実』（広河隆一著／講談社）
『チェルノブイリ診療記』（菅谷昭著／晶文社）

乱丁、落丁本はお取り替えいたします。定価はカバーに表示してあります。

©Nakasuji Jun 2008, Printed In Japan.
ISBN978-4-576-08048-2
http://www.futami.co.jp/